First Edition

D0999088

Motor Vehicle Structures: Concepts and Fundamentals

Dear Professor Cole,

    I thought you might be interested in a book that is authored by one of your former students. I attended U-M in 1976-1977 for my Masters Degree. Your advice and counceling were very helpful to me.

        With very best wishes,

        Stan Serpento

        12-1-01

Mr. Stan T. Serpento
28164 Grand Duke Dr
Farmington Hls, MI 48334-5217

THE LEADERS AND BEST

To Janusz Pawlowski, Guy Tidbury and Roger Masch – three
great engineers of the automobile world, all insistent that
any analysis should start from the fundamentals.

Motor Vehicle Structures: Concepts and Fundamentals

Jason C. Brown, A. John Robertson
Cranfield University, UK

Stan T. Serpento
General Motors Corporation, USA

OXFORD   AUCKLAND   BOSTON   JOHANNESBURG   MELBOURNE   NEW DELHI

Butterworth-Heinemann
Linacre House, Jordan Hill, Oxford OX2 8DP
225 Wildwood Avenue, Woburn, MA 01801-2041
A division of Reed Educational and Professional Publishing Ltd

A member of the Reed Elsevier plc group

First published 2002

© Jason C. Brown, A. John Robertson, Stan T. Serpento 2002

All rights reserved. No part of this publication may be reproduced in
any material form (including photocopying or storing in any medium by
electronic means and whether or not transiently or incidentally to some
other use of this publication) without the written permission of the
copyright holder except in accordance with the provisions of the Copyright
Designs and Patents Act 1988 or under the terms of a licence issued by the
Copyright Licensing Agency Ltd, 90 Tottenham Court Road, London,
England W1P 9HE. Applications for the copyright holder's written permission
to reproduce any part of this publication should be addressed
to the publishers

British Library Cataloguing in Publication Data
A catalogue record for this book is available from the British Library

Library of Congress Cataloguing in Publication Data
A catalogue record for this book is available from the Library of Congress

ISBN 0 7506 5134 2

For information on all Butterworth-Heinemann publications visit our website at www.bh.com

PLANT A TREE
British Trust for Conservation Volunteers
FOR EVERY TITLE THAT WE PUBLISH, BUTTERWORTH-HEINEMANN
WILL PAY FOR BTCV TO PLANT AND CARE FOR A TREE.

Typeset in 10/12pt Times Roman by Laser Words Pvt. Ltd., Chennai, India
Printed and bound in Malta

Contents

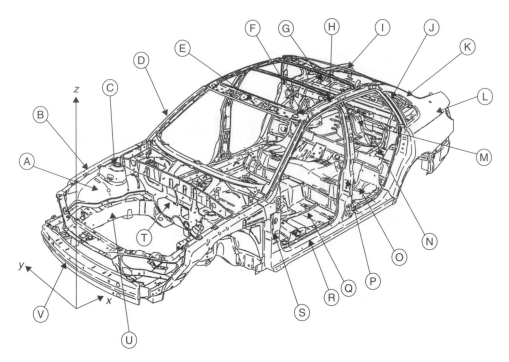

Glossary of 'body-in-white' components (courtesy of General Motors).

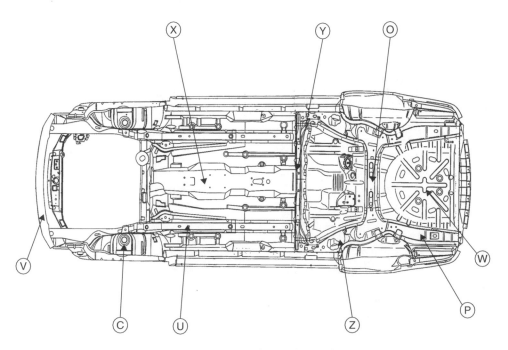

Glossary of underfloor structure components (courtesy of General Motors).

Glossary of 'body-in-white' components

Item	UK description	US description
A	Inner wing panel	Motor compartment side panel
B	Upper wing member	Motor compartment upper rail
C	Suspension tower	Shock tower
D	Upper 'A'-pillar	'A'-pillar or windshield pillar
E	Windscreen header rail	Windshield header or front header
F	Roof stiffener	Roof bow
G	Rear parcel tray	Package shelf
H	Cantrail	Side roof rail
I	Backlight frame	Backlite header or rear header
J	'C'-pillar	'C'-pillar
K	'D'-pillar	'D'-pillar
L	Rear quarter panel	Rear quarter panel
M	Boot floor panel	Rear compartment pan
N	Rear seatback ring	Rear seatback opening frame
O	Rear seat panel	Rear seatback panel
P	'B'-pillar	'B'-pillar or center pillar
Q	Floor panel	Floor pan
R	Sill	Rocker or rocker panel
S	Lower 'A'-pillar	Front body hinge pillar (FBHP)
T	Dash panel	Dash panel
U	Engine (longitudinal) rail	Motor compartment lower rail
V	Front bumper	Front bumper
W	Spare wheel well	Spare tire well
X	Centre (longitudinal) tunnel	Tunnel
Y	Rear seat cross-beam	# 4 crossbar
Z	Rear suspension support beam	# 5 crossbar

Acknowledgements

We would like to acknowledge, with thanks, the following people and organizations for permission to use photographs and diagrams:

Automotive Design Engineer Magazine (Fig. 1.1), Automotive Engineering Magazine (SAE) (Fig. 10.25), Audi UK (Fig. 3.20), Amalgamated Press (Fig. 3.7), Caterham Cars Ltd (Fig. 3.16), Citroen SA (Fig. 3.21), Deutsches Museum Munich (Fig. 3.13), Ford Motor Co. (Fig. 4.13), General Motors Corporation (cover picture, Figs 1.2, 2.1, 2.2, 2.3, 2.4, 3.23, 6.34, 8.1, 8.3, 10.1, 10.4, 10.14, 11.4, 12.2, 12.3), Honda UK and Automotive Engineering Magazine (Fig. 6.35), Lotus Cars Ltd (Figs 3.14, 3.19), Mercedes Benz AG (Fig. 10.27), Motor Industry Research Association (Fig. 2.6), National Motor Museum Beaulieu (Fig. 2.7), Mr Max Nightingale (Fig. 3.18), Oxford University Press (Fig. 3.4), Toyota Motor Corporation (Fig. 4.11), TVR Ltd (Figs 3.15 and 3.17), Vauxhall Heritage Archive, Griffon House, Luton, Bedfordshire, England (Figs 3.2, 3.5 and 5.13), Volkswagen AG (Figs 6.23(a), 6.23(b), 12.1(a), 12.1(b)), The ULSAB Consortium (Fig. 3.25).

Figures 3.10 and 3.22 have been reproduced from the Proceedings of the Institution of Automobile Engineers (Booth, A.G., Factory experimental work and its equipment, *Proc. IAE*, Vol. XXXIII, pp. 503–546, Fig. 25, 1938–9, and Swallow, W. Unification of body and chassis frame, *Proc. IAE*, Vol. XXXIII, pp. 431–451, Fig. 11, 1938–9) by permission of the Council of the Institution of Mechanical Engineers.

Every effort has been taken to obtain permission for the use, in this book, of externally sourced material, and to acknowledge the authors or owners correctly. However, the source of some of the material was obscure or untraceable, and so if the authors or owners of such work require acknowledgement in a future edition of this book, then please contact the publisher.

We would like to thank the academic editor for the series, Prof. D. Crolla, and the editorial staff at Butterworth-Heinemann, particularly Claire Harvey, Sallyann Deans, Rebecca Rue, Renata Corbani, Sian Jones and Matthew Flynn for their efficient, professional and helpful support. Mr Ivan Sears (General Motors Corporation) kindly gave up his valuable time to read our original draft and to suggest many improvements. Mrs Mary Margaret Serpento (Master Librarian) contributed her expertise and time to suggest the method and format for the index. We also received co-operation and help beyond the call of duty from Mr Mike Costin (eminent automotive authority), Mr Dennis Sherar (Archivist, Vauxhall Heritage Centre), Mr Nick Walker (VSCC), and Mrs Angela Walshe (our typist) and we thank them.

We owe a deep debt of gratitude to Dr Ing. Janusz Pawlowski (deceased) and to Mr Guy Tidbury for originating the Simple Structural Surfaces method and for sharing their wisdom and experience with us over the years.

Finally we must thank our wives, Anne, Margaret, and Mary Margaret for their patience, humour and moral support during the writing of the book.

Jason C. Brown
A. John Robertson
Stan T. Serpento

About the authors

Jason C. Brown

Jason Brown had 10 years experience in engineering design and development in the automotive industry, including finite element analysis and vehicle structure and impact tests at Ford Motor Company and stress calculations and vehicle chassis layout and design for various specialist vehicle manufacturers. He has an MSc degree from Cranfield (for which he won the Rootes Prize). Since joining the University staff in 1982, his lecturing, research, and consultancy work has been in testing, simulation and design of automobile structures, vehicle crashworthiness, and non-linear finite element crash-simulation software-development, some of this in co-operation with major companies (including Ford, GM, and others) and with government bodies (British Department of Transport and Australian Federal Office of Road Safety).

A. John Robertson

John Robertson began his engineering career as an engineer apprentice with the de Havilland Aircraft Co. During his apprenticeship he obtained his degree as an external student of the University of London. After working on the design of aircraft controls he moved to Cranfield University to work on vehicle structures. He has developed his interest in overall vehicle concepts and the design of vehicle mechanical components. Recently he has been Course Director for the MSc in Automotive Product Engineering.

Stan T. Serpento

Stan Serpento earned his Bachelors degree in Mechanical Engineering from West Virginia University, and a Masters degree in Mechanical Engineering from the University of Michigan, USA. He began his career at General Motors in 1977 as a summer intern in the Structural Analysis department. Later assignments included vehicle crashworthiness, durability, and noise and vibration work in analysis, development, and validation. Currently he is the Vehicle Performance Development manager for future cars and trucks at the General Motors Global Portfolio Development Center in Warren, Michigan.

Disclaimer

Whilst the contents of this book are believed to be true and accurate at the time of going to press, neither the authors nor the publishers make any representation, express or implied, with regard to the accuracy of the advice and information contained in this book, and they cannot accept any legal responsibility or liability for any errors or omissions that may be made. Neither the authors nor the publisher nor anybody who has been involved in the creation, production or delivery of this book shall be liable for any direct, indirect, consequential or incidental damages arising from the use of information contained in it.

1

Introduction

Objective

- To describe the purpose of this book in the context of a simplified approach to conceptual design.

1.1 Preface

The primary purpose of this book is to demonstrate that the application of a simplified approach can benefit the development of modern passenger car structure design, especially during the conceptual stage. The foundations of the simplified approach are the principles of statics and strength-of-materials that are the core of basic engineering fundamentals. The simple structural surface method (SSS), which originated from the work of Dr Janusz Pawlowski, is offered as a means of organizing the process for rationalizing the basic vehicle body structure load paths. Students will find the approach to be a structured application of the basic engineering fundamental building blocks that are part of their early curricula. Practising engineers may find that a refresher course in statics and strength-of-materials would be helpful. It is hoped that the practice of the simplified approach presented will result in more robust conceptual design alternatives and a better fundamental understanding of structural behaviour that can guide further development.

The category of light vehicle structures described in this book encompasses the many types of passenger car, light trucks and vans. These vehicles are designed and produced with methods and technologies that have evolved over approximately 100 years. In this time the technologies used have become more numerous and also more complex. As a result more staff with a wide range of expertise have been employed in the process of designing and producing these vehicles. The result of this diversification of design methods and production technologies is that an individual engineer rarely has the need to look at the overall design. This book attempts to look at the overall structural design starting at the initial concept of the vehicle.

The initial design of a modern passenger car begins with sketches, moving then to full-size tape drawings and then to three-dimensional clay models. From these models the detail coordinates of the outside shape are finalized. At this stage the 'packaging' of the vehicle is investigated. The term 'packaging' means the determination of the space required for the major components such as the engine, transmission, suspension, steering system, radiator, fuel tank, and not least the space for passengers, luggage or payload. Amongst all these different components and their specialized technologies the vehicle structure must be determined in order to satisfactorily hold the complete vehicle together – the structure is hidden under the attractive shapes determined by style and aerodynamics, does not appear in power and performance specifications, and is not noticed by driver or passenger. Nevertheless it is of paramount importance that the structure performs satisfactorily.

In the majority of cases vehicles are constructed with sheet steel that is formed into intricate shapes by pressing, folding and drawing operations. The parts are then joined together with a variety of welding methods. There are alternative materials such as aluminium and composite materials while other methods of construction include ladder chassis and spaceframes.

Because the structure has to satisfy so many roles and is influenced by so many parameters means that vehicle system designers, production engineers, development engineers as well as structural engineers must be informed about the structural integrity of the vehicle. This book describes a method of structural analysis that requires only limited specialist knowledge. The basic analysis used is limited to the equations of statics and strength of materials. The book therefore is designed for use by concept designers, 'packaging' engineers, component designers/engineers, and structural engineers. Specialists in advanced structural analysis techniques like finite element analysis will also find this relevant as it provides an overall view of the load paths in the vehicle structure.

The method used in this book for studying the load paths in a vehicle structure is the simple structural surfaces (SSS) method. As its name implies compared to modern finite element methods it is a relatively easy method to understand and apply. Professional engineers and university engineering students will find the book applicable to creating vehicle structural concepts and for determining the loads through a vehicle structure.

Although the finite element method (FEM) is mentioned frequently, this book is not intended to treat finite elements in depth. Nor is it the authors' intention to offer the simplified approach as a replacement for finite element analysis (FEA). Rather, the authors suggest the operational potential for FEA to be used in complementary fashion with the simplified approach. A more comprehensive development of the relationship between finite element methods and the simplified conceptual approach is outside the scope of this book.

1.2   Introduction to the simple structural surfaces (SSS) method

The simple structural surfaces method (SSS method) is shown in this book to be a method that is used at the concept stage of the design process or when there are fundamental changes to the structure. The procedure is to model or represent the structure of the vehicle as a number of plane surfaces. Although the modern passenger car, due to aerodynamic and styling requirements has surfaces with high curvature the structure behind the surfaces can be approximated to components or subassemblies that can be represented as plane surfaces.

Each plane surface or simple structural surface (SSS) must be held in equilibrium by a series of forces. These forces will be created by the weight of components attached to them, for example the weight of the engine/transmission on the engine longitudinal rails. The rails are attached to adjacent structural members that provide reactions to maintain equilibrium. The adjacent members therefore have equal and opposite forces acting on them. This procedure of determining the loads on each SSS is continued through the structure from one axle to the other until the overall equilibrium of the structure is achieved. When modelling a structure in this way it can soon be realized if an SSS has insufficient supports or reactions and hence that the structure has a deficiency. Therefore the SSS method is useful for determining that there is continuity for load paths and hence for determining the integrity of the structure.

The authors are not the originators of this method. The SSS method must be credited to the late Dr Janusz Pawlowski of the Warsaw Technical University. Some aspects of this method were first published in the United Kingdom in his book *Vehicle Body Engineering* published by Business Books Limited in 1969. Although based in Warsaw Dr Pawlowski was a frequent visitor to Cranfield University where he developed many of his ideas and where they were passed on to two of the authors.

Dr Pawlowski applied his method to designing passenger coaches (buses) at Cranfield University and in Warsaw he applied it to passenger cars, buses and trams in both academic work and as a consultant to the Polish Automotive Industry.

In addition to the SSS method that forms the basis of this book, additional material by each of the authors has been included. Aspects of the principles of the SSS method applied to local detail design features and examples of real world design as well as academic problems are described.

1.3   Expectations and limitations of the SSS method

No engineering or mathematical model *exactly* represents the real structure. Even the most detailed finite element model has some deficiencies. A model of a structure created with SSSs, like any other model, will not give a complete understanding of how a structure behaves. Therefore, it is important that when using this method the user appreciates what can be understood about a structure and the parameters that cannot be determined.

The SSS method enables the engineer to know the type of loading condition that is applied to each of the main structural members of a vehicle. That is whether the component has bending loads, shear loads, tension loads or compression loads. It enables the nominal magnitudes of the loads to be determined based on static conditions and amplified by dynamic factors if these are known.

One main feature of the method if applied correctly is to ensure that there is continuity for the load path through the structure. It reveals if an SSS has lack of support caused by the omission of a suitable adjacent component. This in turn indicates where the structure will be lacking in stiffness.

When the nominal loads have been determined using basic strength of materials theory the size of suitable components can be determined. However, like any theoretical analysis many practical issues such as manufacturing methods and environmental conditions will determine detail dimensions. The main limitation in the SSS method is that it cannot be used to solve for loads on redundant structures. Redundant structures are constructed in such a way that some individual component's are theoretically not necessary (i.e. parts are redundant). In redundant structures there is more than one load path and the sharing of the loads is a function of the component relative stiffness and geometry. The passenger car structure and other light vehicle structures are highly redundant and therefore it may first be assumed that the method is unsatisfactory for this application. The user of this method must first select a simplified model of the structure and determine the loads on the various components. An alternative, simple model can then be created and the loads determined. The result is that although the exact loads have not been determined the type of load (i.e. shear, bending, etc.) has been found and the detail designer can then design the necessary structural features into the component or subassembly.

When designing vehicle structures it is important to ensure that sufficient stiffness as well as strength are achieved. The SSS method again does not enable stiffness values to be determined. Nevertheless the method does reveal the loads on components such as door frames and this in turn indicates the design features that must be incorporated to provide stiffness.

The user of the SSS method, whether a stylist, component designer, structural analyst or student with an understanding of these expectations and limitations, can gain considerable insight into the function of each major subassembly in the whole vehicle structure.

1.4 Introduction to the conceptual design stage of vehicle body-in-white design

The conceptual phase is very important because it is critical that functional requirements precede the development of detailed design and packaging. With the advent of advanced computer aided design, it is possible to generate design data faster than before. If the structural analyst operates in a sequential rather than concurrent mode, it will be a challenge to keep up with design changes. The design may have been updated by the time a finite element model has been constructed. As a result, the analysis may need to be reworked. In effect, the design process may be thought of as a fast moving train. To intercept this train and steer it on a different track would require that it stop long enough to assess the design's adequacy before it again departs toward its destination. Selecting the right concept is analogous to establishing the correct track and route for the train to follow. This must be done up-front in the process, lest the train be required to take expensive and time consuming detours as its journey progresses.

In the fast paced competitive world, design cannot always wait for sequential feedback. A design–analyse–redesign–reanalyse mode is inconsistent with the demands of today's shortened development times. More concurrent and proactive methodologies must be applied. The conceptual design stage with the integration of computer aided engineering processes has the capability in effect to lay the track (and alternative tracks) that this fast moving train may move on. It must also 'ride along' with the train rather than try to intercept it. The SSS method can provide a tool for rationalizing structural concepts prior to and during the application of CAE tools for certain load conditions. It should be borne in mind, however, that the SSS method is but one of many possible alternative approaches to body structure design. This book is not intended to be a criticism of traditional design methods or CAE or FEA. Rather the book hopes to illustrate how the SSS method would appropriately assist during the conceptual phase. The case studies and guidelines presented in subsequent chapters, which are used to illustrate the SSS method, are examples of many possible alternative design approaches.

Alternative concepts need to be studied within the vehicle's dimensional, packaging, cost and manufacturing constraints *before* the commencement of detail design. Having alternative concepts available and on the shelf to pick and choose from ahead of time is one possible approach. These alternatives may be based on profound knowledge reinforced with existing detailed analysis from previous models and test experience. The analogy is the civil engineers' construction manual that may contain many possible structural cross-sections, joints, etc. from which to choose for a particular application. Another approach is to develop concepts starting with knowledge of basic engineering principles (which comprise the SSS method) and then progress to more tangible representations of the vehicle structure using FEA.

Development of functional requirements for a new body-in-white design should begin with a qualitative free body diagram (FBD) of the fundamental loads acting on the structure, followed by shear and bending moment diagrams for beam members and shear flow for panels using pencil and paper. The same techniques can be used for comparing proposed concepts against existing or current production configurations. It has been the authors' experience that this approach typically results in a higher degree of fundamental understanding of the problem. Consequently, it might be identified early that (1) the structure will need to carry more bending moment in a particular area, or (2) that a particular suspension attachment point will see higher vertical loads because of the movement of a spring or damper, or (3) the elimination of a structural member will now require an alternative load path. It is important that the fundamental issues be identified early, as they may conflict with the original assumptions on which the new product is based. Such a 'first-order' approach should be applied to guide early design proposals and subsequent computer analysis.

The 1950s and especially the early 1960s saw many automotive technical journal articles dealing with the application of fundamental calculations to guide structural design before CAE became commonplace. Often these calculations were applied by experienced designers familiar with engineering fundamentals as well as by engineers with degree qualifications. The engineering and design functions were often identical. Their creativity was evident by the wide variety of automotive body structure design approaches that appeared in the *Automobile Design Engineering* journal (UK) during that period. While construction features from past models may have been utilized, the designs reflected a certain amount of ingenuity and application of basic structural fundamentals. Figure 1.1 shows some examples.

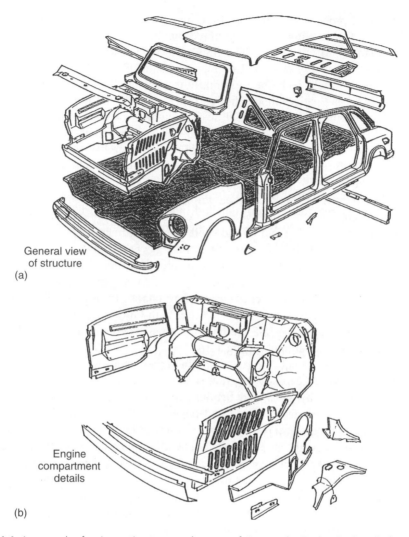

General view
of structure
(a)

Engine
compartment
details

(b)

Figure 1.1 An example of an innovative structure (courtesy of *Automotive Design Engineering*).

1.5   Context of conceptual design stage in vehicle body-in-white design

Each company has their own process for how conceptual design is integrated with the total vehicle design evolution. Conceptual design in this book is defined as the activity that precedes the start of detailed design. The conceptual stage may be performed in conjunction with the preliminary study of alternative platforms upon which to base the design, or in conjunction with the study of model variants off a given platform. The amount of design information that is available to begin a new design is typically much less than the data that exists from an existing platform or current model variant. One of the objectives of conceptual design is to establish the boundaries or limits from which the detailed design can start, especially if the existing platform exerts constraints on the possible design alternatives. Alternative load paths will be considered, as well as overall sizing envelopes for the major structural members. It will be determined which load cases must be addressed now and which will be addressed during the later design phases. Usually there are a few governing load cases that drive the conceptual structure design. These are mainly crashworthiness, overall stiffness (i.e. bending and torsion), and extreme road loading conditions. Questions about the major structural members will be asked such as: 'What *are* the particular governing load cases?', 'How big should the members be sized overall and what are the packaging constraints?', 'Where should load paths be placed?, 'What are the range of materials and thickness to consider?', 'What are the capabilities of alternative platform structures to sustain the loads?', 'What manufacturing processes will be required?' and 'What is the structure likely to weigh?' Issues that concern detailed individual part design thickness, shape, and material grade are left to the later detailed phases unless they are of significant risk to warrant early evaluation.

1.6   Roles of SSS with finite element analysis (FEA) in conceptual design

For a new body-in-white structure in the conceptual stage, alternative load paths and structural member optimization may be studied using relatively coarse finite element models with relatively fast turnaround time when compared to more detailed models used in later stages. An example is the beam–spring–shell finite element model depicted in Figure 1.2.

If the new platform structure must support multiple body types, there may not be sufficient resources to assess all possible variants during a given period of time. Fast methods of assessing the impact of these variants on the base structural platform are

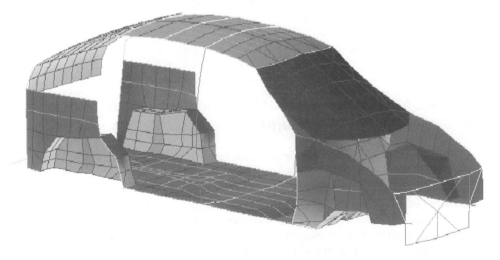

Figure 1.2 Example of preliminary body finite element model during conceptual stage.

desirable. SSS models may be constructed and applied quickly to identify the 'worst case' variant and where to focus the bulk of FEA resources during the conceptual stage.

For an evolutionary body-in-white structure, the primary load carrying members are packaged within an environment that may be constrained by the previous 'parent' design. The evolutionary design is not totally 'new', but rather an established design modified to fit a new package. A structural analysis specialist will recommend the minimum section properties, material characteristics, local reinforcements and joint construction types for the new or non-carryover parts. Preliminary loads are established from a similar existing model until new loads can be generated by test or simulation. The load-path topology is similar to the parent model. Therefore, existing finite element models can be modified and utilized for further study. The role of the SSS method would be for:

- *Qualitative* conceptual design of joints and attachment point modifications.
- Assisting interpretation of the computer aided results and rationalizing load paths. The SSS method is not regarded as an evaluation technique *per se*, but rather as an aid to help rationalize the effect of alternate load paths from a fundamental standpoint.
- Selecting subsequent iterations to be performed on the FEA models for further development.

1.7   Relationship of design concept filtering to FEA models

The SSS method may be regarded as a tool to help *qualitatively* filter design alternatives during the conceptual stage for certain fundamental load cases, and, as mentioned earlier, to help guide the course of FEA iterations during that stage. The combination of these tools can be thought of as laying a foundation for the later design phase, and the more detailed FEA models that follow.

Coarse finite element models act to help filter out and select the concept to be used at the start of detailed design. Larger (more degrees of freedom) finite element models are generally applied in the detailed design phase. However, there may be cases where the application of detailed models during the conceptual stage is appropriate and necessary. Each case will depend on the manufacturer's philosophy, the degree of carryover model vs new model part content for the vehicle body, and the particular issues at hand.

1.8   Outline summary of this book

In this book, Chapter 2 considers the road loads applied to the structure of passenger cars and light goods vehicles. Road loads are caused when the vehicle is stationary, when traversing uneven ground and by the driver when subjecting the vehicle to various manoeuvres. The loads generated when the vehicle is moving are related to the static loads by various dynamic factors. The two main loading conditions are bending, due to the weight of the various components and torsion caused when the vehicle traverses uneven ground. Other loading conditions, due to braking, cornering and when striking pot-holes and kerbs, for example, are also discussed but in less detail.

All the loading conditions that are considered in this book and for which the SSS method is applied of course fall into the category that only cause elastic deformation of the structure. The other category of loads that is not considered here are loads that cause plastic deformation (i.e. impact loads). The subject of crashworthiness is of paramount importance but the mechanism of absorbing energy by plastic deformation is another technology. This is outside the scope of this book and there is more than sufficient material for a separate book on this subject.

Chapter 3 provides a historical overview of car structures. Early chassis frame structures with coach built timber bodies, chassis frames with cruciform bracing led on to chassis with tubular rather than open section rails. Later passenger cars moved on to integral or unitary structures where the chassis and body are combined to give improved strength and stiffness. This type of structure is now almost universal for passenger sedan (saloon) cars constructed in steel sheet. Variations of this type are perimeter frames and alternative materials are aluminium sheet or extrusions. Special vehicles with triangulated tubular spaceframes were sometimes (and still are) used for sports cars. Other special structural concepts such as punt type for sports cars are also illustrated. As the most popular construction is the integral structure this book concentrates on the analysis of this type of structure with some references to the special vehicles.

Having appreciated the loads that are carried and the types of structure used for passenger cars the concept of analysis by the simple structural surfaces (SSS) method is introduced. Chapter 4 details the principles of the method applying it first to simple box-like structures and then with simple models of passenger cars and vans. This chapter concludes with the role that the SSS method can play in a vehicle structural concept.

An example of the method is described in detail in Chapter 5 with the application to a 'standard' sedan (saloon). The equations for the forces on the major components are derived for the four load cases of bending, torsion, cornering and braking. Many vehicle platforms are produced with body variants. A range of a particular model may have sedan, hatchback, estate car, van and pick-up variants. Alternative SSS models for these variants are developed, analysed and discussed in Chapter 6.

Once the loads on a particular SSS or structural subassembly have been obtained the effect on the internal loads or stress needs to be investigated. The load conditions in planar and grillage structures are investigated in Chapter 7. Examples of internal stresses and loads are given for ring structures, sideframes, floors (including normal loading), trusses, and panels.

A case study of a medium size saloon (sedan) car is worked in Chapter 8. The numerical values or the forces on each SSS are evaluated for the bending and torsion load case. The results are illustrated with bending moment and shear force diagrams for all the major components. Alternative models for the front and rear structure are included to illustrate alternative load paths through the structure.

The role of how the SSS method can be used in design synthesis is discussed in Chapter 9. In the design process there are constant changes to the proposed vehicle that can make structural calculations obsolete and necessitate recalculation. By using the SSS method wisely unnecessary reworking can be avoided. This chapter also illustrates the relationship that can be developed with finite element methods and other methods of analysis.

Although the SSS method has been shown to evaluate the loads on major subassemblies the principles of the method can be applied to relatively small subassemblies. Detail case studies are included in Chapter 10 where the method is used in the development of the design of a rear longitudinal rail, a steering column mounting, an engine mounting and a subframe supporting a double wishbone suspension.

Chapter 11 discusses the properties of fabricated sections and spot welded joints. The choice of simplified open or closed sections and their comparison with typical passenger sections are made. Design of spot welded joints for shear, bending and torsional strength is also illustrated. The application of design data sheets is used to determine spot weld pitch, panel buckling and vibration modes.

Finally Chapter 12 shows an industrial view of how the SSS method can be used when applied to the initial design of a body structure. The development of a new vehicle or a variant from an existing platform may need a rapid appraisal of the effect of changed loads or changes in the upper structure. The principles described earlier in the book are shown to be useful for these rapid appraisal procedures.

1.9 Major classes of vehicle loading conditions–running loads and crash loads

As has been noted in the previous sections this book concentrates on the running loads that are applied to the structure. These are the loads that occur in normal service, including extreme conditions of road irregularities and vehicle manoeuvres. The designer using the methods described in this and other books must ensure that the structure is sufficiently strong that no yielding of the material or joint failure results. This means that only loads that cause elastic strains and stresses in the structure are studied.

The main running load cases are the bending of the vehicle due to the weight of the components and/or those due to the symmetrical bump load and the torsion load case. As will be explained, the pure torsion case cannot exist alone, but is always combined with the bending case. By treating these two cases with the principle of superposition the real case of torsion can be analysed. Other cases that will be briefly analysed are the lateral load case due to the vehicle turning a corner and the longitudinal load due to braking.

The reader will probably challenge the authors as to why they have not considered the major subject of crash loads or crashworthiness. It is true today that vehicle designers probably spend more time satisfying a vehicle's crashworthiness than its running loads. The technologies that have to be employed in crashworthiness include dynamics, strain rates and non-linear/non-elastic energy absorption. These are subjects in their own right and can form the basis of several separate books. In order to keep this book to a manageable length the authors decided to limit the work to the running loads. Crashworthiness must remain the subject for a future book or left to the reader to consult the many technical papers published on this subject.

2

Fundamental vehicle loads and their estimation

2.1 Introduction: vehicle loads definition

The principal loads applied to first order and early finite element analysis are gross simplifications of actual complex road loading events.

The actual process begins with sampling of the customer load environment on public roads. These company programs involve instrumenting a statistically valid sample of vehicles and measuring their use in customers' hands across applicable geographic regions. These data are then used to create, modify or update company proving ground road schedules to better match real world customer usage.

The simplified load estimates presented in the following sections are recommended to be applied only in the preliminary design stage, when the absence of test or simulation data warrants it. They should always be qualified and updated as more information becomes available. Additionally, each company will have its own load factors, based on experience of successful designs, which may not necessarily be identical to the load factors presented in this book.

2.2 Vehicle operating conditions and proving ground tests

Environment and customer usage data are the historical basis for the road surface types, test distance, speed and number of repetitions applied to the proving ground's durability test schedules.

The proving ground can provide the equivalent of, for example, 100 000 miles (160 000 km) of high severity customer usage in a fraction of the distance. Because even this fraction can represent several months or more of actual testing, companies have developed laboratory tests to further compress development and validation time. These tests provide a simulation of the proving ground's road load environment through computer programmed actuators applied at the tyres or wheel spindles.

As far as the passenger car body structure is concerned, the significant proving ground events can be reduced to two types:

(a) instantaneous overloads;
(b) fatigue damage.

Table 2.1 compares these load types and lists some of the typical proving ground events. Figures 2.1 and 2.2 illustrate some of these events. A collection of road load durability events that a vehicle is tested for is called a schedule. The list of events which make up a schedule such as potholes, Belgian blocks, etc. are called subschedules.

In addition there are transport loads which occur when the vehicle is being shipped from the factory to the retail agency. These may occur from overseas, truck, air, and rail transport. These loads are typically simplified in calculation by a static force vector applied as a percentage of the gross vehicle weight. The tie-down attachment location

Table 2.1

Type of load	Number of event repetitions	Load amplitude (N)	Acceptance criteria	Proving ground event examples
Instantaneous overload	Low: <10	High: 10^4	Limited permanent deformation, maintenance of function	Large pot-holes, kerb bumps, large bumps, panic braking, high g cornering, high power-train torque, overland transport, service
Fatigue	High: 10^2	Moderate: 10^3	Cycles or distance to crack initiation, limited crack propagation, maintenance of function	Cobblestone track, medium size pot-holes, Belgium block road, twist course, transport, service

Figure 2.1 Example of fatigue loading event.

Figure 2.2 Example of proving ground event.

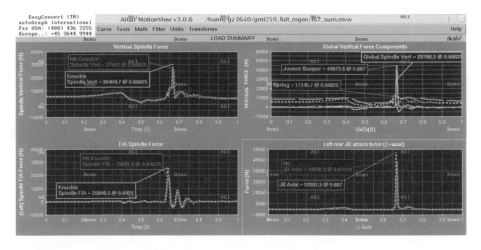

Figure 2.3 Example of transient load trace.

and method of lashing have a strong influence on the reaction loads into the body structure.

Service loads are those which occur when the vehicle must be serviced either by the customer or a technician. Examples are jacking to change a tyre, towing, hoisting, or retrieval of a disabled vehicle from a ditch. There are usually location points designated to help ensure that the structure is not damaged from improper use.

Figures 2.3 and 2.4 describe different loading types. Instantaneous overloads are characterized by short duration transient events with high amplitudes.

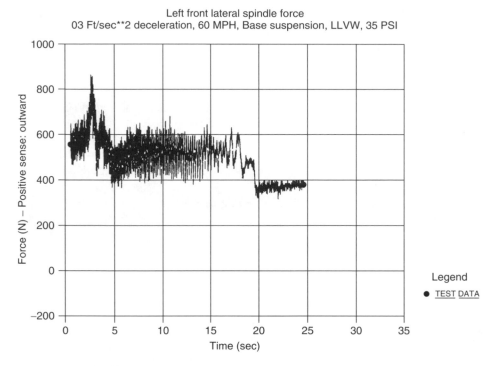

Figure 2.4 Example of fatigue load trace.

Fatigue loads are characterized by complex time histories with lower amplitudes but a greater number of occurrences. These time histories, when multiplied by the number of occurrences or block repetitions of the event, constitute a number of cycles on the order of 10^4 to 10^5.

Raw time histories are usually condensed and processed by rainflow cycle counting or other techniques to derive the number of cycles at each stress level. This data is then used to perform structural life estimations.

The next sections of this chapter represent gross simplifications of the dynamic load spectra into static load estimates which are used for first order structural synthesis and analysis.

2.3   Load cases and load factors

The vehicle designer needs to know the worst or most damaging loads to which the structure is likely to be subjected, (a) to ensure that the structure will not fail in service due to instantaneous overload and (b) to ensure a satisfactory fatigue life.

At the very early design stage (as covered in this book), the main interest focuses on instantaneous strength. The considerable attention (through test and analysis) which is paid to fatigue life is outside the scope of this book. A commonly used assumption at the early design stage is: 'If the structure can resist the (rare) worst possible loading which can be encountered, then it is likely to have sufficient fatigue strength'.

For early design calculations, the actual dynamic loading on the vehicle is often replaced by a 'factored static loading', thus:

dynamic load ≡ (static load) × (dynamic load factor)

An extra 'factor of safety' is sometimes used:

i.e.   equivalent load ≡ (static load) × (dynamic load factor) × (safety factor)

In order to apply this approach, certain load cases are considered. For early design consideration, these will be 'global' road load cases, i.e. affecting the structure as a whole. As the design develops, local load cases (e.g. door slam, hinge loads, bracket forces, etc.) will be used.

Crash cases are often the most difficult and critical to design for. They are outside the scope of this book, since the structure moves out of the elastic regime into deep collapse. However, the support forces for an energy absorbing part of the vehicle could form a static load case.

2.4   Basic global load cases

The principal 'normal running' global road load cases are as follows (see Figure 2.5 for axis directions):

1. *Vertical symmetrical* ('bending case') causes bending about the Y–Y axis
2. *Vertical asymmetric* ('torsion case') causes torsion about the X–X axis and bending about the Y–Y axis.
3. *Fore and aft* loads (braking, acceleration, obstacles, towing)
4. *Lateral* (cornering, nudging kerb, etc.)
5. Local load cases, e.g. door slam, etc. ⎫ Not considered here
6. Crash cases   ⎭

The load cases and load factors used vary from company to company, but some typical values, or ways of estimating them, are listed below.

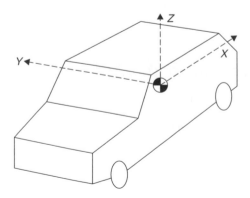

Figure 2.5 Vehicle axis system.

2.4.1   Vertical symmetric ('bending') load case

This occurs when both wheels on one axle of the vehicle encounter a symmetrical bump simultaneously (see Figure 2.6). This applies a bending moment to the vehicle about a lateral axis.

Some values for dynamic factor and additional safety factor from different workers are listed in Table 2.2.

For off road vehicles, dynamic factors up to 6 have been used.

2.4.2   Vertical asymmetric case (and the pure torsion analysis case)

This occurs when only one wheel on an axle strikes a bump. An extreme example of this is shown in Figure 2.7. Vertical asymmetric loading applies torsion as well as bending to the vehicle body. It has been found that torsion is a more severe case to design for than bending.

Figure 2.6 Vertical symmetric load case (courtesy of MIRA UK).

Table 2.2 Bending load factors for cars

	Commonly used	Erz (1957)	Pawlowski (1969)
Dynamic factor	3	2	2.5
Additional safety factor	1.5		1.4–1.6 (away from stress concentrations) 1.5–2.0 (engine and suspension mountings)

Figure 2.7 Vertical asymmetric load case (courtesy of National Motor Museum, Beaulieu).

Different vehicles will experience different torsional loads, for a given bump height, depending on their mechanical and geometric characteristics. In order to relate the torsion loading of any vehicle to *operating conditions*, Erz (1957) suggested that the asymmetric loading should be specified by the *maximum height H* of a bump upon which one wheel of one axle rests, with all other wheels on level ground.

The torque so generated will depend on the roll stiffnesses of the front and rear suspensions and on the torsion stiffness of the vehicle body. These act as three torsion springs in series, thus the overall torsional stiffness K_{TOTAL} is given by:

$$\frac{1}{K_{\text{TOTAL}}} = \frac{1}{K_{\text{FRONT}}} + \frac{1}{K_{\text{BODY}}} + \frac{1}{K_{\text{REAR}}}$$

where K_{FRONT} and K_{REAR} are the roll stiffnesses of the front and rear suspensions and K_{BODY} is the torsional stiffness of the body (i.e. about a longitudinal axis).

The vehicle body is usually much stiffer about the longitudinal axis than the front and rear suspensions. Thus its contribution to the overall twist θ is often negligible. In such cases the term $1/K_{\text{BODY}}$ is small and can be omitted from the above equation.

Thus, using the notation from Figure 2.8, the torque T generated by bump height H (all wheels in contact) is given by:

$$T = K_{\text{TOTAL}}\theta$$

But the twist at axle 1,

$$\theta \approx H/B \quad \therefore \quad T = K_{\text{TOTAL}}\frac{H}{B} \tag{2.1}$$

The torque T is caused by weight transfer onto the wheel on the bump from the wheel on the other side of the axle.

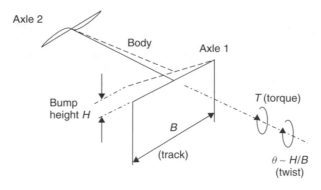

Figure 2.8 Torque generated by bump height H.

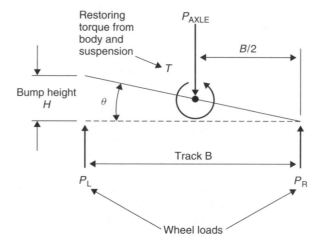

Figure 2.9 Forces and moments on axle 1.

Using the notation in Figure 2.9, this may be shown as follows:
For vertical force equilibrium at axle 1:

$$P_L + P_R = P_{AXLE}$$

i.e.
$$P_L = P_{AXLE} - P_R \qquad (2.2)$$

where P_{AXLE} is the total axle load, and P_L and P_R are the left- and right-hand wheel reactions.

For moment equilibrium:
$$T = (P_L - P_R)\frac{B}{2}$$

substituting from (2.2):
$$T = (P_{AXLE} - 2P_R)\frac{B}{2} \qquad (2.3a)$$

$$P_R = P_{AXLE}/2 - (T/B) \qquad (2.3b)$$

similarly:
$$P_L = P_{AXLE}/2 + (T/B) \qquad (2.3c)$$

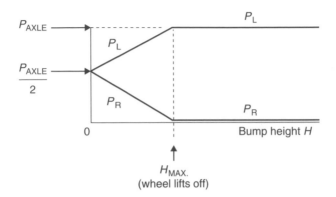

Figure 2.10 Static wheel reactions for bump under left wheel.

The torque will reach a limit when wheel R lifts off, i.e. when $P_R = 0$ (and hence $P_L = P_{AXLE}$). From (2.3b)/(c), the wheel loads are seen to behave as shown in Figure 2.10.
   Note it is always a wheel on the *lightest loaded axle* which lifts off:

i.e.
$$T_{MAX.} = P_{AXLE}\frac{B}{2} \tag{2.4}$$

where $P_{AXLE} =$ load on lightest loaded axle.
   Thus the maximum torque $T_{MAX.}$ in this limiting case may be obtained from (2.1) and (2.4)

$$T_{MAX.} = K_{TOTAL}\frac{H_{MAX.}}{B} = P_{AXLE}\frac{B}{2} \tag{2.5}$$

where H_{MAX} is the bump height to cause wheel R to lift off.

Thus:
$$H_{MAX.} = \frac{P_{AXLE}B^2}{2K_{TOTAL}} \tag{2.6}$$

   Often, for modern passenger cars with soft springs, the suspension will strike the 'bump stops' for asymmetric bumps smaller than H_{MAX}. The torsion load would then be applied to the vehicle through the bump stop (much stiffer than the suspension spring).
   Different workers have suggested different values of H for the torsion case. Some of these values are given in Table 2.3. Pawlowski (1969) suggested that an extra dynamic factor be applied if the vehicle will frequently encounter rough conditions (e.g. pot-holed ice).

Table 2.3 Torsion bump height for cars

	Pawlowski (1969)	Erz (1957)
Bump height	0.2 m	0.2 m
Inertial factor	1.3	
Inertial factor (off road)	1.8	

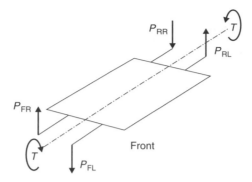

Figure 2.11 Pure Torsion Load Case.

Pure torsion load case

Erz (1957) and Pawlowski (1969) both suggested that the torque generated by this load case should be applied as a *pure torsion* load case. For this, the bending component of the vertical asymmetric case is removed, leaving equal and opposite pure couples at either end of the vehicle (see Figure 2.11). This could not occur in practice since it would require negative wheel reactions. However, the pure torsion load case is important because it generates very different internal loads in the vehicle structure from those in the bending load case, and, as such, is a different structural design case. This will be discussed in Chapters 4 and 5. The total loads in the vehicle structure can be calculated by superimposing the separate results of the calculations for pure torsion and pure bending, etc. This is explained in section 2.5.

2.4.3   Longitudinal loads

(a) Clutch-drop (or snap-clutch) loads

The longitudinal accelerations from this case have been found to be smaller than braking loads, except for towing, when a factor of 1.5 has been used (but this applies a special loading to the car). However, this case is quite severe on drive-train mounting interfaces. It can also result in high vertical/opposite loads on mounting fixes.

(b) Braking

Table 2.4 shows overall braking load factors suggested by various workers.

Since the braking forces at the ground contact patches are offset by a vertical distance h from the vehicle centre of gravity, there will be weight transfer from the rear to the front wheels.

Table 2.4 Load factors for braking

	Pawlowski (1969)	Cranfield tests (Tidbury 1966)	Garrett (1953)
Cars	1.1 g	1.84 g	1.75 g
Trucks	0.75 g		

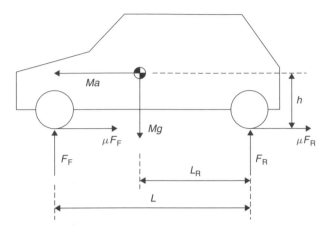

Figure 2.12 Weight transfer in braking.

Using the notation in Figure 2.12, for longitudinal force equilibrium:

$$Ma = \mu Mg = \mu F_F + \mu F_R \tag{2.7}$$

where μ = tyre friction coefficient, M = vehicle mass, a = braking deceleration and Mg = vehicle weight.

For moment equilibrium about the rear contact patch:

$$F_F L = Mg L_R + Mah \tag{2.8}$$

Thus, from (2.7) and (2.8) the front axle vertical reaction F_F is:

$$F_F = Mg(L_R + \mu h)/L$$

Similarly, the rear axle vertical reaction F_R is:

$$F_R = Mg(L_F - \mu h)/L$$

(c) Longitudinal load on striking a bump

Using the notation from Figure 2.13 and assuming static equilibrium, in which case the resultant wheel reaction passes through the wheel centre:

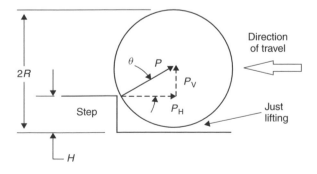

Figure 2.13 Longitudinal bump case.

Vertical equilibrium: $P \sin \theta = P_V$

Horizontal equilibrium: $P \cos \theta = P_H$

Thus: $P_H = (P_V / \sin \theta) \cos \theta = (P_V / \tan \theta)$

where P_v = static vertical wheel load and P_H is the horizontal force developed, and

$$\sin \theta = (R - H)/R = 1 - (H/R)$$

(Assuming approximately equal rolling and free radii of the tyre.)

This neglects dynamic effects including wheel inertia. These are very important in this case and Garrett (1953) suggested a dynamic load factor $K_{DYN} = 4.5$ so that:

$$P_H = K_{DYN}(P_V / \tan \theta)$$

For a given bump height H and vertical wheel force P_V, the horizontal force P_H depends on wheel radius (smaller wheels developing larger forces) as illustrated in Figure 2.14. At large step sizes approaching the magnitude of the wheel radius, the longitudinal force becomes very large, because the term $\tan \theta$ approaches zero. In reality, the longitudinal force could not reach infinity, as shown in the table, because the strength of the suspension would set a limit on the forces experienced by the vehicle.

Pawlowski (1969) suggested the step height H should be the same as for the torsion (vertical asymmetric) case.

H/R	P_H/P_V (static)
0.1	0.48
0.25	0.88
0.293	1.00
0.5	1.73
0.75	3.87
1.0	∞

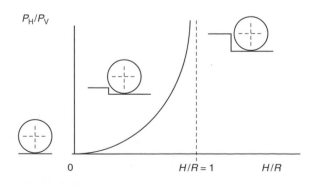

Figure 2.14 Longitudinal load plotted against height of step (no dynamic factor).

2.4.4   Lateral loads

Lateral loads on the vehicle can be limited by a number of situations.

(a) Sliding of tyres (cornering, see Figure 2.15(a) and 2.16)
    max. force $= \mu\, Mg$

where $Mg =$ vehicle weight and $\mu =$ friction coefficient (see Table 2.4).

(b) Kerb nudge ('overturning')
The lateral force reaches a maximum when the wheel (A) opposite the kerb just lifts off. (Actual rollover of the car will not occur unless there is sufficient energy before impact to lift the vehicle centre of gravity to point B above the kerb contact point C after impact). Using the symbols in Figure 2.15(b) and taking moments about point C:

$$F_{(\text{LAT})}h = Mg\frac{B}{2} \times K$$

$$F_{(\text{LAT})} = \frac{MgB}{2h} \times K$$

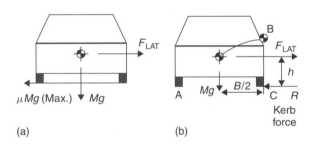

(a)   (b)

Figure 2.15 Lateral load cases.

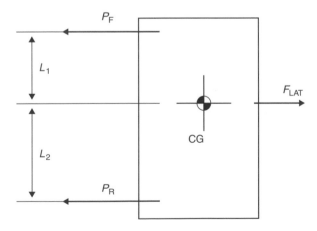

Figure 2.16 Fore and aft distribution of wheel loads in cornering.

where $F_{(LAT)}$ = lateral force, h = height of CG above ground, B = track, K = dynamic safety factor (due to short duration pulse and rotational inertia). Garrett (1953) suggested $K = 1.75$. Other workers suggested $K = 1.4$.

It should be borne in mind that the above formulae are *not* intended as a means of predicting or assessing a rollover event. Rather, it is intended for making a preliminary estimate of the lateral force acting on the vehicle body and attachments.

(c) Fore and aft distribution of lateral loads

As a rough first approximation for fore/aft lateral load distribution, take moments about the vertical axis through the centre of gravity of the vehicle (this assumes static equilibrium in yaw). Using the nomenclature in Figure 2.16:

$$P_F L_1 - P_R L_2 = 0 \quad \text{(moments about the centre of gravity)}$$

but
$$P_F + P_R = F_{LAT} \quad \text{(lateral force equilibrium)}$$

Combining and rearranging:

$$P_F = \frac{F_{(LAT)} L_2}{(L_1 + L_2)}$$

$$P_R = \frac{F_{(LAT)} L_1}{L_1 + L_2} \quad (F_{(LAT)} = \text{total lateral force, e.g. from turning})$$

As with the longitudinal cases, in the lateral cases there will be weight transfer onto the outer wheels due to the offset h between the ground and the vehicle centre of gravity.

2.5   Combinations of load cases

For the sake of calculation, the load cases are usually split into separate idealized cases and the results are then combined, by addition (i.e. using the principle of superposition) to give the effect of the real loads. The main idealized load cases are: (1) bending (as in symmetrical vertical load case); (2) pure torsion; (3) the lateral cases; (4) the longitudinal cases.

Example   Vertical asymmetric case

Consider an internal load in one of the structural members (e.g. shear force Q, see Chapters 4 and 5). If the bending load case causes it to take value Q_B and the pure torsion case causes it to take value Q_T then in the combined case its value will be $Q_B + Q_T$. The combination of loads on the vehicle is illustrated in Figure 2.17.

In the wheel lift-off case (see section 2.4.2) the pure torsion case wheel load on the lightest loaded axle would be equal to the wheel load for the pure bending case on that axle. Suppose, for example, that the lightest loaded axle were at the rear. Then for the wheel lift-off case P_2 would be equal to P_R. Then the resultant rear wheel loads would become $2P_R$ on the left, and zero on the right. On the other axle, at the front in this example, the wheel loads would be $(P_F - P_T)$ and $(P_F + P_T)$ on the left- and right-hand sides respectively.

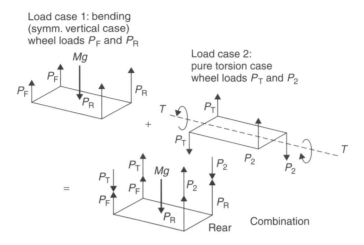

Figure 2.17 Vertical asymmetric loading as a combination of cases.

2.5.1   Road loads

The vertical asymmetric case illustrates another point, namely that in real life, many of the pure load cases described in section 2.4 will occur in combination with each other. For example, the vehicle weight is always present, so that the longitudinal cases (section 2.4.3) and the lateral cases (section 2.4.4) are always accompanied by, at least, a 1 *g* vertical load (and sometimes more, for example the vertical symmetrical bump case (section 2.4.1)).

On the road, almost any combination of the pure cases from section 2.4 can be encountered. For example, a cornering vehicle might encounter a bump (or pot-hole) with one wheel. This could involve the extreme vertical asymmetric load case (section 2.4.2) plus the extreme factored longitudinal bump case (section 2.4.3(c)) plus the cornering case (section 2.4.4(a)) and possibly also the braking case (section 2.4.3(b))!

Some of the cases cannot occur together in their extreme forms. Thus, tyre adhesion limitations mean that the full straight line braking and full pure cornering loads cannot occur simultaneously, and allowance must be made for this.

3

Terminology and overview of vehicle structure types

3.1 Basic requirements of stiffness and strength

The purpose of the structure is to maintain the shape of the vehicle and to support the various loads applied to it. The structure usually accounts for a large proportion of the development and manufacturing cost in a new vehicle programme, and many different structural concepts are available to the designer. It is essential that the best one is chosen to ensure acceptable structural performance within other design constraints such as cost, volume and method of production, product application, etc.

Assessments of the performance of a vehicle structure are related to its *strength* and *stiffness*. A design aim is to achieve sufficient levels of these with as little mass as possible. Other criteria, such as crash performance, are not discussed here.

3.1.1 Strength

The strength requirement implies that no part of the structure will lose its function when it is subjected to road loads as described in Chapter 2. Loss of function may be caused by instantaneous overloads due to extreme load cases, or by material fatigue. Instantaneous failure may be caused by (a) overstressing of components beyond the elastic limit, or (b) by buckling of items in compression or shear, or (c) by failure of joints. The life to initiation of fatigue cracks is highly dependent on design details, and can only be assessed when a detailed knowledge of the component is available. For this reason assessment of fatigue strength is usually deferred until after the conceptual design stage.

The strength may be alternatively defined as the maximum force which the structure can withstand (see Figure 3.1). Different load cases cause different local component loads, but the structure must have sufficient strength for all load cases.

3.1.2 Stiffness

The stiffness K of the structure relates the deflection Δ produced when load P is applied, i.e. $P = K\Delta$. It applies only to structures in the elastic range and is the slope of the load vs deflection graph (see Figure 3.1).

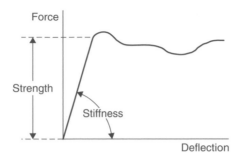

Figure 3.1 The concepts of stiffness and strength.

The stiffness of a vehicle structure has important influences on its handling and vibrational behaviour. It is important to ensure that deflections due to extreme loads are not so large as to impair the function of the vehicle, for example so that the doors will not close, or suspension geometry is altered. Low stiffness can lead to unacceptable vibrations, such as 'scuttle shake'.

Again, different load cases require different stiffness definitions, and some of these are often used as 'benchmarks' of vehicle structural performance. The two most commonly used in this way are:

(a) *Bending stiffness* K_B, which relates the symmetrical vertical deflection of a point near the centre of the wheelbase to multiples of the total static loads on the vehicle. A simplified version of this is to relate the deflection to a single, symmetrically applied load near the centre of the wheelbase.

(b) *Torsion stiffness* K_T, relates the torsional deflection θ of the structure to an applied pure torque T about the longitudinal axis of the vehicle. The vehicle is subjected to the 'pure torsion load case' described in section 3.4 (where the torque is applied as equal and opposite couples acting on suspension mounting points at the front and rear), and the twist θ is measured between the front and rear suspension mountings. Twist at intermediate points along the wheelbase is sometimes also measured in order to highlight regions of the structure needing stiffening.

These two cases apply completely different local loads to individual components within the vehicle. It is usually found that the torsion case is the most difficult to design for, so that the torsion stiffness is often used as a 'benchmark' to indicate the effectiveness of the vehicle structure.

3.1.3   Vibrational behaviour

The global vibrational characteristics of a vehicle are related to both its stiffness and mass distribution. The frequencies of the global bending and torsional vibration modes are commonly used as benchmarks for vehicle structural performance. These are not discussed in this book. However, bending and torsion stiffness K_B and K_T influence the vibrational behaviour of the structure, particularly its first natural frequency.

3.1.4   Selection of vehicle type and concept

In order to achieve a satisfactory structure, the following must be selected:

(a) The most appropriate structural type for the intended application.
(b) The correct layout of structural elements to ensure satisfactory load paths, without discontinuities, through the vehicle structure.
(c) Appropriate sizing of panels and sections, and good detail design of joints.

An assumption made in this book is that if satisfactory load paths (i.e. if equilibrium of edge forces between simple structural surfaces) are achieved, then the vehicle is likely to have the foundation for sufficient structural (and especially torsion) stiffness. Estimates of interface loads between major body components calculated by the simplified methods described are assumed to be sufficiently accurate for conceptual design, although structural members comprising load paths must still be sized appropriately for satisfactory results. Early estimates of stiffness can be obtained using the finite element method, but the results should be treated with caution because of simplifications in the idealization of the structure at this stage.

3.2   History and overview of vehicle structure types

Many different types of structure have been used in passenger cars over the years. This brief overview is not intended to be a detailed history of these, but to set a context. It covers only a selection of historical and modern structures to show the engineering factors which led to the adoption of the integral structure for mass produced vehicles, and other types for specialist vehicles.

3.2.1   History: the underfloor chassis frame

In the 1920s, when mass production had become well established, the standard car configuration was the separate 'body-on-chassis' construction. This had certain advantages, including manufacturing flexibility, allowing different body styles to be incorporated easily, and allowing the 'chassis' to be treated as a separate unit, incorporating all of the mechanical components. The shape of the chassis frame was ideally suited for mounting the semi-elliptic spring on the beam axle suspension system, which was universal at that time. Also, this arrangement was favoured because the industry at that time was divided into separate 'chassis' and 'body' manufacturers. Tradition played an additional part in the choice of this construction method.

The underfloor chassis frame, which was regarded as the structure of the car, consisted of a more or less flat 'ladder frame' (Figure 3.2). This incorporated two open section (usually pressed C-section) sideframes running the full length of the vehicle, connected together by open section cross-members running laterally and riveted to the side frames at 90° joints. Such a frame belongs to a class of structures called '*grillages*'.

A grillage is a flat ('planar') structure subjected to loads normal to its plane (see Figure 3.3). The *active* internal loads in an individual member of such a frame are (see inset):

Figure 3.2 Open section ladder frame chassis of the 1920s (courtesy of Vauxhall Archive Centre).

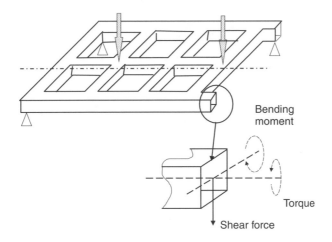

Figure 3.3 Grillage frame.

(a) Bending about the in-plane lateral axis of the member.
(b) Torsion about the longitudinal axis of the member in the plane of the frame.
(c) Shear force in a direction normal to the plane of the frame.

Open section members, as used in 1920s and 1930s chassis frames, are particularly flexible locally in torsion. Further, the rivetted T-joints were poor at transferring bending moments from the ends of members into torsion in the attached members and vice versa. Chassis frames from that era thus had very low torsion stiffness. Since, on rough roads, torsion is a very important loading, this situation was not very satisfactory. The depth of the 'structure' was limited to a shallow frame underneath the body, so that the bending stiffness was also relatively low.

Texts from the 1920s show that considerable design attention was paid only to the *bending* behaviour of the structure, mainly from the strength point of view.

The diagram in Figure 3.4 (from Donkin's 1925 textbook on vehicle design) shows carefully drawn shear force and bending moment diagrams for the chassis frame, based on the static weight of the chassis, attached components, body, payload, etc. The bending moment diagram is compared with the distribution of bending strength in the chassis side members. Important to note, however, is the complete absence of

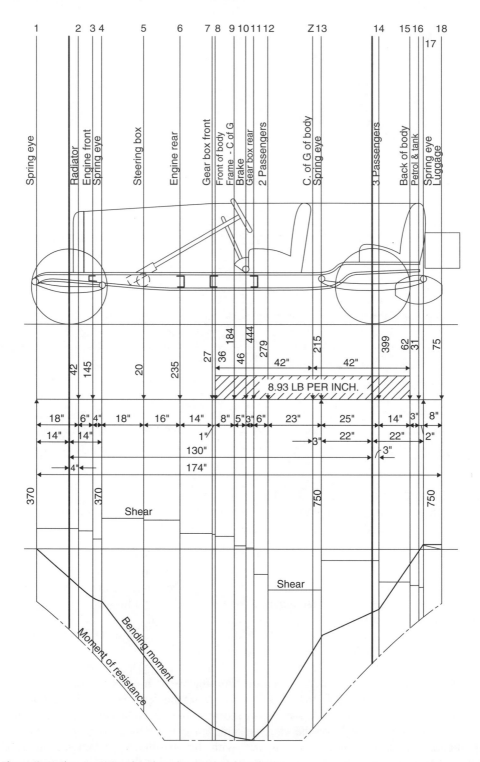

Figure 3.4 Chassis structural analysis diagram (Donkin 1925).

any consideration of *torsion* behaviour of the structure. The importance of this was not fully understood by the engineering community until later.

Good torsional design is important to ensure satisfactory vehicle handling, to avoid undesirable vibrations, and to prevent problems of incompatibility between body and frame as described below. The torsion load case (see Chapter 2) puts different local loads on the structural components from those experienced in the bending load case. Torsion stiffness is often used as one of the 'benchmarks' of the structural competence of a vehicle structure.

In view of the poor torsion performance of the early chassis frame, it is perhaps fortunate that car bodies in the 1920s (Figure 3.5) were 'coachbuilt' by carpenters, out of timber, leading to body structures of very low stiffness. In the early part of the 1920s, the majority of passenger cars had open bodies which, as we will see later in the book, are intrinsically flexible. At that time, it was commonly *assumed* that the body carried none of the road loads (only self-weight of body, passengers and payload), and consequently it was not designed to be load bearing. This was particularly true for torsion loads.

Early experience with metal clad bodies, particularly in 'sedan' form (i.e. with a roof), where torsion stiffness was built in fortuitously and inadvertently, led to problems of 'rattling' between the chassis and the body, and also 'squeaking' and cracking at various points within the body which were, unintentionally, carrying structural loads.

The root of these problems lay in the fact that the 'body-on-chassis' arrangement consists, in essence, of two structures (the body and the chassis) acting as torsion springs in *parallel*.

For springs in parallel, the load is shared between the springs in proportion to their relative stiffnesses. This is a classic case of a 'redundant' or 'statically indeterminate' structural system. In the simplified case where the body and chassis are connected only

Figure 3.5 Car body manufacture in the 1920s (courtesy of Vauxhall Archive Centre).

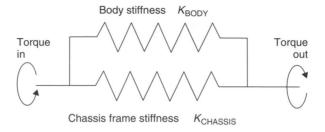

Figure 3.6 Springs in parallel.

at their ends (as in Figure 3.6):

$$T_{\text{TOTAL}} = T_{\text{BODY}} + T_{\text{CHASSIS}}$$

$$K_{\text{TOTAL}} = K_{\text{BODY}} + K_{\text{CHASSIS}}$$

$$T_{\text{BODY}}/T_{\text{CHASSIS}} = K_{\text{BODY}}/K_{\text{CHASSIS}}$$

where T = torque and K = torsional stiffness.

Thus, in the case of a flexible body on a (relatively) stiff chassis frame, most of the torsion load would pass through the chassis. Conversely, if the body were stiff and the chassis flexible, then the body would carry a larger proportion of the torsion load.

As the 1920s progressed, this was implicitly recognized (based on practical experience) in the construction approach. Bodies were deliberately made flexible (particularly in torsion) by the use of flexible metal joints between the resilient timber body members, and by deliberate use of flexible materials for the outer skin of the body (Figure 3.7).

It will be seen later in the book that closed shell structures such as closed car bodies are very effective in torsion, with the outer skin subjected locally to shear. In order to keep torsion stiffness (and hence loads) small, flexible materials such as fabric were

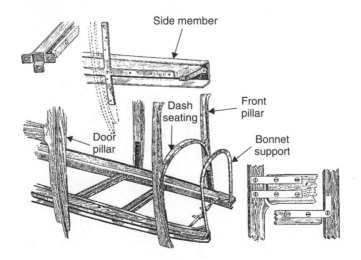

Figure 3.7 Construction details of a timber framed 'fabric body'.

used to form the outer skin of the body. In Europe, the 'Weymann fabric saloon' body was a well-known, and much copied, example of this.

An alternative approach was to use very thin aluminium cladding with deliberate structural discontinuities at key points to relieve the build-up of undesired stresses in the body.

The fabric-covered wood-framed car body was not amenable to large-scale mass production. As the 1920s gave way to the 1930s, the requirements of high volume production led to the widespread use of pressed steel car body technology. The bodies were formed out of steel sheets, stamped into shape, and welded or riveted together. This led to much greater stiffness in the body, particularly for torsion, because the steel panels were quite effective locally in shear. The overall configuration still remained as the 'body-on-separate- chassis'.

In the 1930s the subfloor chassis frame was still made of open section members, riveted together, and it was still regarded as the 'structure' of the vehicle.

From the 'springs in parallel' analogy, however, it can be seen that a much greater proportion of the load was now taken through the body, owing to its greater stiffness. This led to problems of 'fighting' between the body and the chassis frame (i.e. rattling, or damage to body mounts caused by undesired load transfer between the body and the chassis). Several approaches were tried to overcome this problem. These were used both individually and in combination with each other. They included:

(a) Flexible (elastomer) mountings were added between the chassis frame and the body. Laterally spaced pairs of these mountings acted as torsion springs about the longitudinal axis of the vehicle between the chassis and the body.

Consider a pair of body mounts positioned on either side of the body. If the linear stiffness of the individual elastomer body mounts is K_{LIN}, and they are separated laterally by body width B (see Figure 3.8), then the torsional stiffness K_{MOUNT} of the pair of mounts about the vehicle longitudinal axis is:

$$K_{\text{MOUNT}} = K_{\text{LIN}} B^2 / 2$$

As used in 1930s vehicles this, in effect, made the 'load path' through the body more flexible, because the pairs of soft elastomer mounts and the body formed a chain of

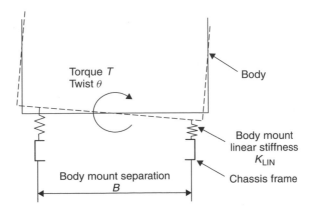

Figure 3.8 Laterally positioned pair of body mountings.

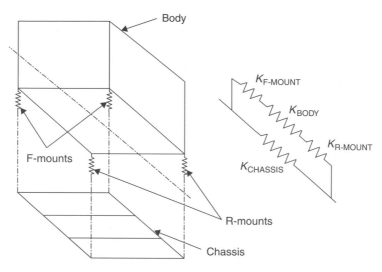

Figure 3.9 Body and body mounts in series in torsion.

'springs in *series*' (see Figure 3.9). For such a system, the overall flexibility (the reciprocal of stiffness) is the sum of the individual flexibilities. Thus, the overall stiffness is lower than that of the individual elements in the series.

Thus:
$$1/K_{\text{TOTAL}} = 1/K_{\text{F-MOUNT}} + 1/K_{\text{BODY}} + 1/K_{\text{R-MOUNT}}$$

where K = torsion stiffness and $1/K$ = torsional flexibility.

Since the body-plus-mountings assembly was, structurally, still in parallel with the chassis, the effect of the reduced stiffness was to reduce the proportion of the torsion load carried by the body.

(b) The converse of the approach in (a) was to stiffen the chassis frame, thus encouraging it to carry more of the load. Open section, riveted chassis frame technology was still the norm in the 1930s, and so a method of increasing torsion stiffness, but still using open section members, was needed.

A common solution was the use of *cruciform bracing*. For this, a cross-shaped brace, made usually of open channel section members, was incorporated into the chassis frame as shown in Figure 3.10. It was necessary for the ends of this to be well connected, in shear, to the chassis side members.

Figure 3.11 shows how the cruciform brace works as a torsion structure. On the left, the 'input torque' is fed in as a couple consisting of two equal and opposite forces. This couple, or torque, is reacted by the couple composed of the equal and opposite forces on the right-hand side of the diagram.

The exploded view shows the local loads in the individual members. Member A has vertical loads downward at both ends. These are reacted by the upward force at point C. This in turn is reacted by an equal and opposite force at point C on member B, and the loads in member B are a mirror image of those in A.

Although the *overall* effect is a torsion carrying structure, the *individual members* (A and B) are subject only to bending and shear forces. Hence it was possible to

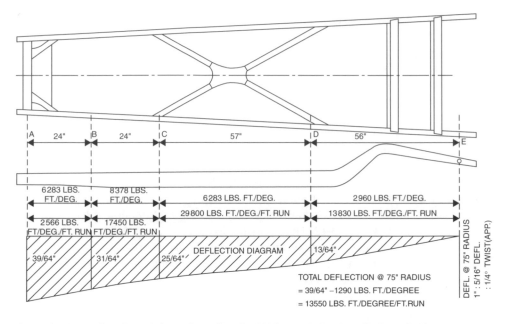

Figure 3.10 Cruciform braced chassis frame (Booth 1938 by permission Council of I.Mech.E.).

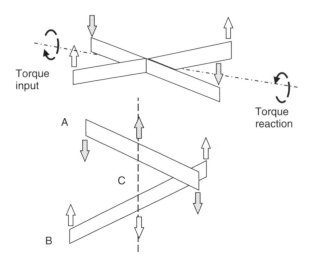

Figure 3.11 Free body diagrams of cruciform brace members.

use open sections. It is essential that there is a good, continuous, bending load path in both members A and B at point C (point of maximum bending moment).

By the mid-1930s, the need for reasonable torsion stiffness was well recognized. A paper of the time (Booth 1938), from which Figure 3.10 is taken, gives overall torsion stiffness values between 1000 and 1750 Nm/deg. for various cruciform-braced chassis frames. One of the best of the cruciform braced underfloor open section chassis was that of the Lagonda V12 of the late 1930s. The frame consisted entirely of a substantial cross brace, with only small tie rods along the side to prevent 'wracking' distortion. Its torsion stiffness was measured to be a little over 2000 Nm/deg. (Bastow 1945 and 1978).

(d) Another way of improving chassis frame torsion stiffness was to incorporate *closed* (i.e. 'box section') cross-members. Closed section members are much stiffer in torsion than equivalent open section ones. Small closed section cross-members, as used in 1930s cars, whilst giving a considerable increase compared with previous chassis (which were extremely flexible in torsion), still gave overall results which would be considered low today.

(e) By the mid-1930s it was realized that the steel body was much stiffer than the chassis in both bending and torsion. Greater 'integration' of the body with the chassis frame was also used in some designs. A body attached to the chassis frame by a large number of screws, thus using some of the high torsion stiffness of the body, is shown in Figure 3.12. This approach led eventually to the modern 'integral body' which is the major topic of the rest of this book.

(f) The ultimate version of the underfloor chassis frame using small section members is the 'twin tube' or 'multi tube' frame.

This is still essentially a ladder type grillage frame, with side members connected by lateral cross-members. However, now both the side members and the cross-members consist of closed section tubes.

The advantage of this arrangement is that typically, for members of similar cross-sectional dimensions, a member with a closed section will be thousands of times stiffer in torsion than an equivalent open section member. Also adjacent members

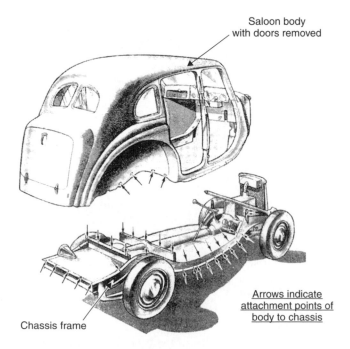

Saloon body
with doors removed

Arrows indicate
attachment points of
body to chassis

Chassis frame

Figure 3.12 Multiple attachments of body to frame.

Figure 3.13 Twin tube frame of Auto Union racing car (1934–1937) (courtesy of Deutsches Museum, München).

are usually welded together in this construction method, and such joints are much stiffer than riveted connections. The overall torsion stiffness of the chassis assembly improves accordingly.

This type of structure was used in specialist racing vehicles between the late 1930s and the 1950s. For example, the Auto Union racing car (1934–1937) shown in Figure 3.13 used this system, as did many sports racing cars in the 1950s, such as Ferrari and Lister Jaguar. Examples of these may be seen in Costin and Phipps' book (1961).

Separate ladder frames, with either open or closed section members, are still used widely on certain types of passenger car such as 'sport utility vehicles'' (SUVs) and they are almost universal on commercial heavy goods vehicles.

3.2.2   Modern structure types

In more modern times, the closed tube (or closed box) torsion structure has been used to greater effect by using larger section, but thinner walled members. The torsion constant J for a thin walled closed section member is proportional to the *square* of the area A_E *enclosed* by the walls of the section.

$$J = 4A_E^2 t/S \quad \text{for a closed section with constant thickness walls.}$$

where t = wall thickness and S = distance around section perimeter.

Hence there is a great advantage in increasing the breadth and depth of the member. Additionally, a large depth will give good (i.e. stiff and strong) bending properties. The torsion *stiffness* K of the closed section backbone member is then:

$$K = GJ/L$$

where G = material shear modulus and L = length of member, so that:

$$T = K\theta$$

where T = applied torque and θ = torsional deflection (twist)).

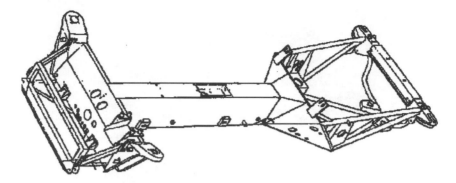

Figure 3.14 Sheet steel backbone chassis (courtesy Lotus Cars Ltd).

The development of better road holding, coming from a better understanding of suspension geometry, made greater body stiffness essential. This, and the push towards welded, pressed sheet steel body technology, led to the widespread use of the 'large section tube' concept in car structures in the post-World War II era. Some examples follow.

Backbone structure

The 'Backbone' chassis structure is a relatively modern example of the 'large section tube' concept (although Tatra vehicles of the 1930s had backbone structures). This is used on specialist sports cars such as the Lotus shown in Figure 3.14. It still amounts to a 'separate chassis frame'.

The backbone chassis derives its stiffness from the large cross-sectional enclosed area of the 'backbone' member. A typical size might be around 200 mm × 150 mm.

It will be seen later that, in tubular structures in torsion, the walls of the tube are in shear. Thus, in the case of the Lotus, the walls of the tube consist of shear panels. However, shear panels are not the only way of carrying in-plane shear loads. For example, a triangulated 'bay' of welded or brazed small tubes can also form a very effective and weight efficient shear carrying structure. It is possible to build an analogue of the 'backbone' chassis frame using triangulated small section tubes This approach is used in some specialist sports cars, such as the TVR shown in Figure 3.15.

Such specialist vehicles often have bodies made of glass reinforced plastic. On many vehicles of this type, the combined torsion stiffness of the chassis and the attached body together is greater than the sum of the stiffnesses of the individual items. This reflects the fact that the connection between the two is not merely at the ends, as discussed earlier, but is made at many points, giving a combined structure which is highly statically indeterminate.

Triangulated tube structure

The triangulated tube arrangement is not limited to backbone structures. Perhaps a more common approach using this principle, particularly for sports cars, is the 'bathtub'

Figure 3.15 Backbone chassis made of triangulated tubes (from author's collection, courtesy TVR Ltd).

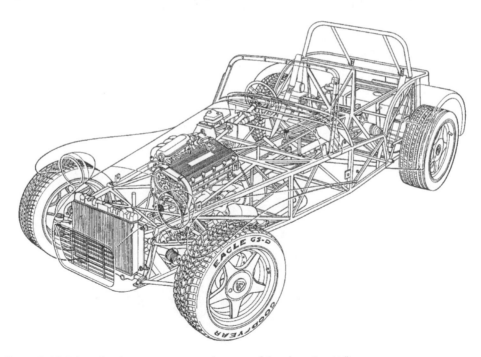

Figure 3.16 Triangulated sports car structure (courtesy of Caterham Cars Ltd).

layout, in which the triangulated structure surrounds the outside of the body. A classic example is the Caterham shown in Figure 3.16. This approach has the advantage that the coachwork can consist of thin sheet metal cladding, attached directly to the frame-work. If the vehicle is an open car, the large cockpit interrupts the 'closed box' needed for torsion stiffness. In such a case, torsion stiffness is sometimes restored by 'boxing in' either the transmission tunnel, or the dash/cowl area (or both) with triangulated bays. Stiffening of the edges of the passenger compartment top opening, particularly at the corners, can also bring some improvement to the torsional performance. The principles of the simple structural surfaces method (see later in Chapters 4 and 5) can still be applied to this type of structure if the triangulated bays (including edges) are treated as structural surfaces.

This method of construction is best suited to low volume production because of low tooling costs. It is not well suited to mass production due to complication and labour intensive manufacture.

Incorporation of roll cage into structure

The ultimate way of using the 'tube' principle is to make the tube encompass the whole car body. A version of this is shown in the triangulated sports car structure shown in Figure 3.17. The triangulated 'roll cage' now extends around the passenger compartment. The enclosed cross-sectional area of the body is thus very large, and hence the torsion constant is large also.

Roots *et al.* (1995) studied the torsion stiffness of a racing vehicle in which the roll cage was incorporated into the structure in a range of different ways. They showed that the torsion stiffness can be increased by over 500 per cent as compared with the basic chassis frame. The contribution of the roll cage depended on (a) the degree of triangulation in the roll cage, and (b) on how well the roll cage was connected to the rest of the structure (i.e. on the continuity of load paths).

Pure monocoque

The logical conclusion of the 'closed box' approach is the 'monocoque' (French: 'single shell'). For this, the outer skin performs the dual role of the body surface and structure.

Figure 3.17 Triangulated sports car structure with integrated roll cage (Roots *et al.* 1995 with original permission from TVR Ltd).

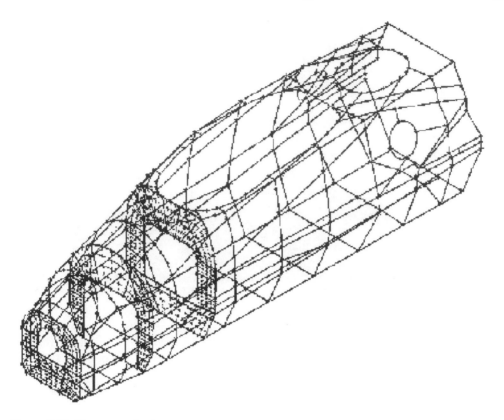

Figure 3.18 Monocoque structure.

This is the automotive version of aircraft 'stressed skin' construction. It is very weight efficient.

The *pure* monocoque *car* structure is relatively rare. Its widespread use is restricted to racing cars as shown in Figure 3.18. There are several reasons for this.

First, for the monocoque to work efficiently, it requires a totally *closed* tube. However, practical vehicles require openings for passenger entry, outward visibility, etc. This requires interruptions to the 'single shell' which then reduce it to an open section, with consequent lowering of torsion stiffness. Also, the shell requires reinforcement: (a) to prevent buckling (stringers or sandwich skin construction often provide this), and (b) to carry out-of-plane loads, e.g. from the suspension (internal bulkheads are often used for this).

Typical Formula 1 racing car monocoques, usually made of carbon fibre composite sandwich material, can have torsion stiffness greater than 30 000 Nm/deg. for the composite 'tub' alone. In such vehicles, the engine and gearbox also act as load bearing structures, in series with the monocoque.

Punt or platform structure

Other modern car chassis types include the 'punt structure'. This is usually of sheet metal construction, in which the floor members (rocker, cross-members, etc.) are of

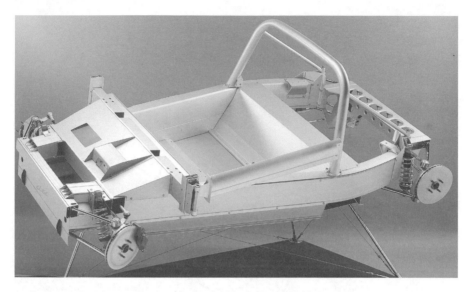

Figure 3.19 Punt chassis (courtesy of Lotus Cars Ltd).

large closed section, with good joints between members. It is thus a grillage structure of members with high torsion and bending properties locally. In many cases (but not all), the upper body is treated as structurally insignificant. The punt structure is often used for low production volume vehicles, for which different body styles, or rapid model changes are required.

The Lotus Elise (aluminium, see Figure 3.19) is an example of a punt structure.

This approach is often also used to create cabriolet or convertible versions of mass produced integral sedan car structures (see Chapter 6, section 6.4).

Perimeter space frame or 'birdcage' frame

Another modern structure is the perimeter or 'birdcage' frame. A typical example is the Audi A2 aluminum vehicle (Figure 3.20).

In this type of structure, relatively small section tubular members are built into stiff jointed 'ring-beam' bays, welded together at joints or 'nodes'. We shall see later that ring beams are moderately effective at carrying local in-plane shear. For this, the edge members of each ring frame, and especially the corners, must be stiff locally in bending.

This choice of construction method is usually dictated by production requirements. In the case of the A2, the various beam sections are of extruded or cast aluminium (with some additional members of pressed sheet), and so they must be assembled into this structural concept using welded 'nodes' or joints.

The individual open-bay ring frame is not a very weight efficient shear structure (see Chapter 7). If the (very high) shear stiffness of the skin panels is incorporated into this type of body, it now becomes an 'integral' structure (see later), and a considerable increase in torsional rigidity is usually observed, depending on the stiffness of the attachment.

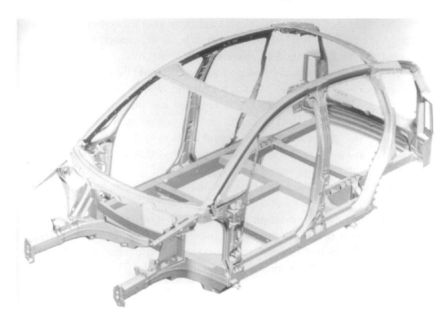

Figure 3.20 Perimeter space frame (courtesy of Audi UK Ltd).

Integral or unitary body structure

The subject of this book, and the most widely used modern car structure type, is the 'integral' (or 'unitary'), spot welded, pressed steel sheet metal body. It is well suited to mass production methods. The body is self-supporting, so that the separate 'chassis' is omitted, with a saving in weight.

The first mass produced true integral car bodies were introduced in the 1930s. A notable example was the Citroën 11 CV, shown in Figure 3.21, which was in production from 1934 to 1956.

Figure 3.21 Citroën 11 CV of 1934 (courtesy of Automobiles Citroën SA).

An interesting study by Swallow (1938) was made on another vehicle, the Hillman 10 HP, when the structure was changed from separate body-on-chassis (1938 model year) to integral construction (1939 model), the vehicles being otherwise largely identical. Amongst other things, the torsional behaviour was compared (Figure 3.22). The torsion stiffness rose from 934 Nm/deg. (689 lbft/deg.) for the chassis and body (together) for the earlier model to 3390 Nm/deg. (2500 lbft/deg.) for the integral body.

An interesting 'hysteresis' effect is visible in the unloading curve for the separate frame vehicle, due to slippage between body and chassis at the mounting bolts. This is absent in the integral structure. Swallow also mentioned that spring rates for axle and engine suspensions, and for the suspension bushes had to be reduced due to increased 'harshness' (the passing of transient forces to the vehicle occupants) in the integral vehicle.

Figure 3.23 shows a modern example of an integral 'body-in-white' (i.e. bare body shell). The integral body is really a mixture of the monocoque and the 'birdcage' types. The body forms a 'closed box' torsion structure (with consequent high stiffness). The walls, or 'surfaces' of the box, consist of the skin panels (such as the roof, floor,

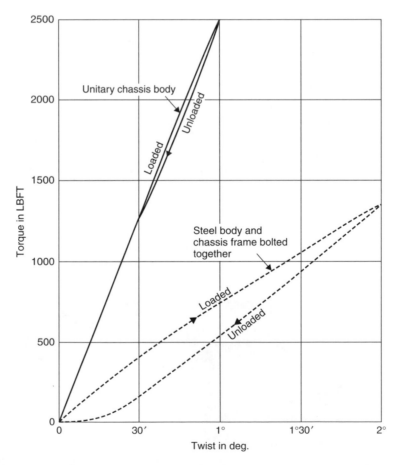

Figure 3.22 Torsion stiffness tests, 1938 and 1939 (Swallow 1938 by permission Council of I.Mech.E.).

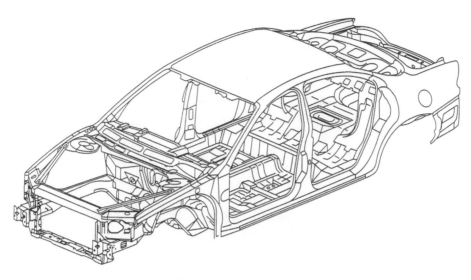

Figure 3.23 Modern integral body-in-white (courtesy General Motors Corporation).

bulkheads, etc.) where possible. Elsewhere open bay ring frames (sideframe, wind-shield frames etc.) form the surface of the box, wherever openings are required. Beam members are also used to carry out-of-plane loads, for example in the floor.

In the integral structure the panels and body components are stamped from sheet steel and fixed together mostly by spot welding, although clinching, laser welding or other methods are sometimes used for particular locations. The beam members are formed out of folded or pressed sheet steel shapes, welded together as shown in Figure 3.24 and in Figure 11.4 (Chapter 11). These beams can be independent (e.g. B-pillar), or they can be formed as part of the larger panels (as in the case of the transmission tunnel), or they can be attached to panels by spot welding (e.g. floor cross-members,

Figure 3.24 Integral body floor assembly, showing structural members.

Figure 3.25 The ultralight steel auto body (ULSAB) (courtesy of the ULSAB consortium).

rockers, etc.). To avoid ugly 'sink marks', attached beam members are never spot welded to externally visible skin panels.

The ultralight steel auto steel body, ULSAB (see AISI 1998) is a modernized version of this theme which may well show the way to near-future developments. In this, hydroforming (the creation of complex cross-sections by forcing tubes into moulds by internal hydraulic pressure) is used widely as an alternative method of forming beam-like components. 'Laser welded blanks' (i.e. of tailored varying thickness) are also used widely, in addition to steel sandwich panels. Laser welding and adhesive bonding, both of which are stiffer than spot welding, are used extensively to join the panels together. The result is a structure which was recorded to be lighter and stiffer than the 'traditional' integral steel bodies it was compared to.

Although, at the time of writing, steel is used almost universally for high volume mass produced car bodies, the suppliers of competing materials, such as aluminium and composite plastics, have been developing integral body technologies also. For example, the aluminium intensive vehicle, AIV, is made of pressed sheet aluminium panels, 'weld-bonded' together. Nardini and Seeds (1989) made a discussion of design issues for aluminium integral bodies.

If properly constructed, the integral body is well capable of carrying torsion, bending and other loads. Because the structure comprises the outer surface of the body, it is much stiffer than most other vehicle structure types, for the reasons given above. Typical values of torsion stiffness for modern integral car bodies are approximately 8000 to 10 000 Nm/deg. for 'everyday' sedans and higher (around 12 000 to 15 000 Nm/deg. or more) for luxury vehicles.

Ideally the in-plane stiffness of the individual panels or surfaces must be used. The precise way in which this occurs in a well-designed integral car body is the subject of the remainder of this book.

4

Introduction to the simple structural surfaces (SSS) method

Objectives

- To introduce the concept of modelling a vehicle as a series of simple structural surfaces (SSS) and to show its usefulness.
- To understand the definition and limitations of a simple structural surface.
- To illustrate the SSS method on a simplified box or van structure.
- To show examples of SSS models representing vehicle structures.

4.1 Definition of a simple structural surface (SSS)

A simple structural surface (SSS) is a plane structural element or subassembly that can be considered as rigid only in its own plane. Figure 4.1 shows such a structural element where the length a and height b are large in comparison to its thickness t.

Considering sections through the element in the $x-y$ plane and $y-z$ plane the second moments of area can be obtained by the standard formulae:

$$I_x = at^3/12$$

$$I_y = tb^3/12$$

$$I_z = bt^3/12$$

As t is small the second moments of area I_x and I_z will be very small compared with I_y that is:

$$I_y \gg I_x$$

$$I_y \gg I_z$$

Therefore the SSS is capable of resisting bending moments about the y-axis, but has little or no resistance for moments about the x- and z-axes.

Direct loads F_x and F_z acting in the plane of the SSS will also be satisfactorily resisted but normal local loads along the y-axis will of course result in flexing of the SSS by bending about the z- or x-axes.

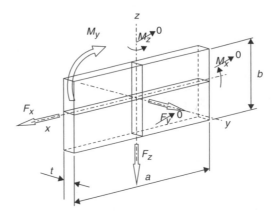

Figure 4.1 Definition of a simple structural surface.

4.2   Structural subassemblies that can be represented by a simple structural surface (SSS)

Passenger car structures consist of a number of subassemblies that can be represented by SSSs. Although due to modern styling and aerodynamic requirements the structure has considerably curved surfaces it will be assumed as a first approximation that these surfaces can be represented by a plane surface. Structural subassemblies that can be represented by an SSS are those that have good rigidity across their whole plane.

Figure 4.2 shows examples of effective SSSs. All of the examples are suitable for carrying shear loads as indicated by the edge loads Q_1 and Q_2. The basic panel

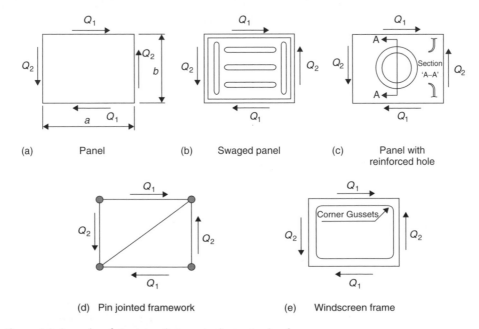

Figure 4.2 Examples of structures that are simple structural surfaces.

shown at (a) will have limited capacity if its thickness is small due to the tendency to buckle. Stiffening of the panel by swaging (b) or a reinforced hole (c) can increase the load capacity. The pin-jointed framework shown at (d) will also provide suitable structural properties for the loads Q_1 and Q_2. A ring frame such as the windscreen frame (e) provided it has sufficient corner joint stiffness and sidebeam stiffness will provide a satisfactory SSS.

The simple panel shown at Figure 4.2(a) can be made more effective by adding stiffeners (booms) along the edges as shown in Figure 4.3(a). In this example the boom/panel assembly is used as a cantilever supported at the right-hand side with a vertical load F_z at the left-hand end. The vertical boom distributes the load F_z into the panel by the edge shear load Q_2. The boom may be attached to the panel by spot welds, rivets, bolts or even made integral with the panel by folding or pressing operations. The edge shear loads Q_1 and end loads K_1 and K_2 are necessary for maintaining equilibrium (see next section).

Figure 4.3(b) illustrates the use of an auxiliary beam to carry loads that are normal to the surface. The vertical force F_z acts normal to the floor panel so the beam shown shaded has been added in the $y-z$ plane that is held in equilibrium by the forces K_3 and K_4. The floor panel itself can of course carry edge shear loads Q_3 and Q_4 in its own $x-y$ plane.

Figure 4.4 illustrates three planar systems that cannot be considered as SSSs. At (a) there is a four-bar linkage mechanism while at (b) and (c) there is respectively a discontinuous ring and a panel with a large cut-out. The structures at (b) and (c) will suffer local bending and large deflections, as illustrated, within their planes.

In contrast to these unsatisfactory SSSs, Figure 4.5 shows a passenger car sideframe and a bus sideframe that form very effective SSSs, provided the corner joints are capable of resisting bending moments.

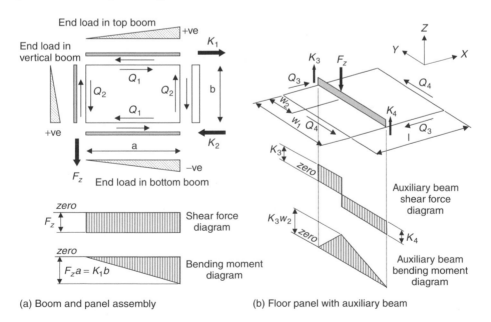

(a) Boom and panel assembly   (b) Floor panel with auxiliary beam

Figure 4.3 Further examples of simple structural surfaces.

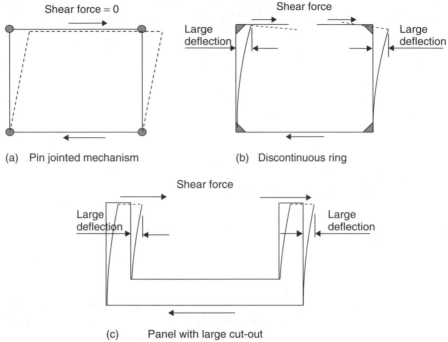

(a)   Pin jointed mechanism   (b)   Discontinuous ring

(c)   Panel with large cut-out

Figure 4.4 Examples of planar systems that are not simple structural surfaces.

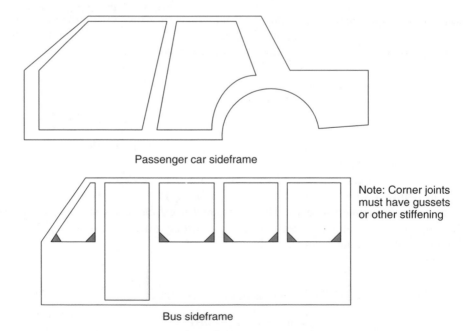

Passenger car sideframe

Bus sideframe

Note: Corner joints must have gussets or other stiffening

Figure 4.5 Examples of vehicle structural assemblies that can be represented as simple structural surfaces.

4.3   Equilibrium conditions

The SSSs illustrated in Figures 4.2 and 4.3 can all be evaluated by the equations of statics to determine the relationship and values of the forces. All the examples of Figure 4.2 are loaded with complementary shear forces. For the panel at (a) take moments about one corner and obtain:

$$Q_1 b - Q_2 a = 0 \qquad (4.1)$$

Therefore, if Q_1 is known Q_2 can be obtained provided the lengths a and b are known. A similar equation can be obtained for all the other examples of Figure 4.2.

The boom/panel assembly of Figure 4.3(a) requires four equations as there are four unknown forces for a known applied load F_z:
Resolving vertically

$$F_Z - Q_2 = 0 \qquad (4.2)$$

Resolving horizontally

$$K_1 - K_2 = 0 \qquad (4.3)$$

Moments about lower right-hand corner

$$F_z a - K_1 b = 0 \qquad (4.4)$$

Resolving horizontally for the top boom

$$Q_1 - K_1 = 0 \qquad (4.5)$$

Equations (4.2), (4.3) and (4.4) are the basic equations of statics giving the three loads required to hold the assembly in equilibrium. Equation (4.5) is the equilibrium equation for the top boom. The equilibrium equation for the lower boom is:

$$Q_1 - K_2 = 0 \qquad (4.6)$$

and for the panel:

$$Q_1 b - Q_2 a = 0 \qquad (4.7)$$

The shear force and bending moment diagram for the assembly are shown and the end loads in the booms indicated.

When SSSs are loaded normal to their plane auxiliary SSSs may be introduced as shown in Figure 4.3(b). Here the floor panel has the addition of a beam, an auxiliary SSS, to carry the applied load F_z. The floor panel itself can carry shear as indicated by the edge loads Q_3 and Q_4 which are complementary shear forces.

Taking moments about a corner of the floor panel:

$$Q_3 w_1 - Q_4 \ell = 0 \qquad (4.8)$$

Taking moments about K_4 at the end of the auxiliary SSS:

$$K_3 w_1 - F_z (w_1 - w_2) = 0 \qquad (4.9)$$

Resolving vertically:

$$K_3 + K_4 - F_z = 0 \qquad (4.10)$$

With these equations all the forces on the auxiliary SSS and the panel can be obtained. The shear force and bending moment diagrams are shown for the auxiliary SSS in Figure 4.3(b).

4.4   A simple box structure

The most simple vehicle type structure is an International Standards Organization (ISO) freight container or box van consisting of six SSSs. Figure 4.6 shows a structure of this type but includes an extra SSS, a cross-beam, that is required to distribute the payload across the floor. Alternative floor structures are discussed in the Appendix at the end of this chapter.

The payload (F_{pl}) plus an allowance for the structure weight (F_i) is applied in the centre of the cross-beam that is supported at the sides of the box by the sidewalls. The forces K_1 which support the cross beam, are then reacted by equal and opposite forces in the sidewalls. It should be noted that the forces K_1 act in the planes of both the cross-beam and the sidewalls and therefore meet the requirement that forces must act in the plane of an SSS. By resolving forces and by symmetry the loads K_1 are equal to half the applied load at the centre of the cross-beam and the following equation is obtained:

$$K_1 = F_{zs}/2 \tag{4.11}$$

The cross-beam is shown positioned at distance 'a' from the front end and the box van is of length L. The sidewall must be held in equilibrium by forces acting at the front and rear edges where these can be reacted in the planes of the front and rear panels.

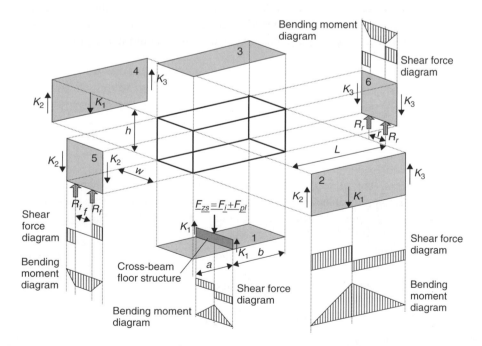

Figure 4.6 Box van with bending load.

Therefore by taking moments about the front lower corner:

$$K_3 = K_1 a / L \qquad (4.12)$$

where K_3 is the edge force between the sidewall and the rear panel.

Resolving forces vertically for the sidewall gives the equation:

$$K_2 + K_3 - K_1 = 0 \qquad (4.13)$$

where K_2 is the force between the sidewall and the front panel.

The front and rear panels are supported by the suspension systems and as the structure is loaded symmetrically resolving forces for the front panel:

$$R_f = K_2 \quad \text{where } R_f \text{ is the reaction from one front suspension}$$

and resolving forces for the rear panel:

$$R_r = K_3 \quad \text{where } R_r \text{ is the reaction from one rear suspension.}$$

Figure 4.6 also shows the shear force and bending moment diagrams for the floor cross-beam, the sidewall and the front and rear panels. Each of these members is subject to shear and bending loads which are functions of their lengths. The floor and roof panels have no loads acting in their planes.

The same simple box van structure is shown in Figure 4.7 when it is subject to torsion. In section 2.4.2, Chapter 2, it was explained that by taking the lighter loaded axle and applying the suspension load upwards on one side and downwards on the other side the maximum static torsion condition is obtained. In this example of Figure 4.7 if $b > a$ then the rear end loads are the lighter and so the lighter reactions R_r are applied as shown. R_r is the reaction from one rear suspension under static bending conditions

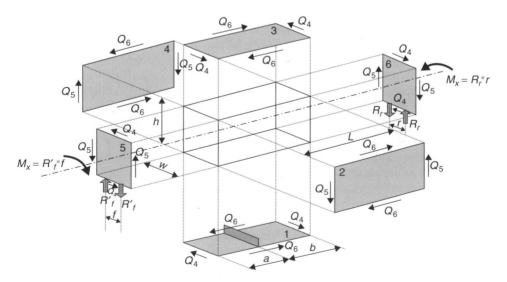

Figure 4.7 Box van in torsion.

as explained previously. For equilibrium of the body the torsion moment M_x at front and rear must be equal therefore:

$$R_f^* f = R_r^* r \tag{4.14}$$

As the lateral spacing of the front and rear suspension mountings 'f' and 'r' respectively may not be equal, the reaction R_f' will not necessarily be equal to R_r nor equal to R_f obtained for the bending case.

At the front panel the moment applied through the suspension mountings is balanced by the edge loads from the sides, floor and roof. Taking moments for the front panel:

$$R_f f - Q_5 w - Q_4 h = 0 \tag{4.15}$$

It should be noted that the front panel has the clockwise 'external' couple M_x equal to $R_f' f$ (as viewed from the front) applied through the suspension mounting points. The edge loads Q_4 and Q_5 all act in an anticlockwise sense. The direction of the arrows is such that the head of each arrow points to the tail of the next arrow.

The equilibrium of the floor is now considered. An equal and opposite force to the front panel force Q_4 acts on the front edge of the floor. To prevent the floor moving laterally there must be an equal and opposite force Q_4 acting at the rear edge. These two forces are separated by the length L producing a moment $Q_4 L$ that must be balanced. Therefore edge loads Q_6 must be applied at each side separated by the distance 'w'. Taking moments about one corner, the moment equation is:

$$Q_4 L - Q_6 w = 0 \tag{4.16}$$

This floor panel is therefore loaded in complementary shear and the directions of the arrows representing the edge loads at the front and right side are such that the heads of the arrows point towards the same corner. Similarly the arrows representing the forces at the rear and left side point at the same corner. Panels that are subject to complementary shear do not have 'externally' applied moments.

The left-hand sidewall has an equal and opposite force Q_6 to that on the floor acting at its lower edge. Similarly it has force Q_5 acting at the front vertical edge that is equal and opposite to the force on the front panel. By resolving forces longitudinally and vertically for equilibrium there must be a force Q_6 acting on the top edge in the opposite direction to that on the lower edge and a force Q_5 acting on the rear edge of the sidewall in the opposite direction to that on the front edge.

The moments generated by the spacing of these forces must be balanced and therefore the moment equation for the left-hand sidewall (moments about one corner) is:

$$Q_6 h - Q_5 L = 0 \tag{4.17}$$

The left-hand sidewall is loaded in complementary shear with no 'external' moment and with the heads of the arrows representing the directions of the edge loads point at the front/lower and rear/upper corners.

The roof and right-hand sidewall have similar but opposite loading conditions to the floor and left-hand sidewall respectively, while the rear panel has similar but opposite edge loads to those on the floor, sidewalls and roof. Note that the arrows of the edge loads on the rear panel act in a clockwise sense as viewed from the front. This must be correct as the moment due to the reaction from the rear suspension forces R_r must act in an anticlockwise sense to balance the clockwise moment applied from the front suspension.

Examination of the three equations (4.15–4.17) shows that there are three unknowns Q_4, Q_5 and Q_6. Therefore solutions may be found:

$$Q_4 = R_f f / 2h \tag{4.18}$$

$$Q_5 = R_f f / 2\omega \tag{4.19}$$

$$Q_6 = R_f f L / 2h\omega \tag{4.20}$$

In practice, a box van must have doors, windows and usually a large door at the rear. If the rear panel of the model is removed as in Figure 4.8 it must be replaced by a ring structure or door frame. Provided the frame has stiff side, top and bottom members and corner joints that resist bending moments (like the windscreen frame illustrated in Figure 4.2(e)) the structure will still resist torsion loads as shown in Figure 4.7. If the rear door frame is omitted or the frame has effectively pin jointed corners and low stiffness edge members, the rear panel cannot carry shear. Therefore we have the load condition shown in Figure 4.8, the edge load Q_4 cannot exist because the missing sides of the door frame and the flexible corner joints cannot transfer loads to the top. No shear loads can be transferred across the door aperture from the suspensions attached to the door sill. As Q_4 does not exist there is no shear force applied to the roof and Q_6 is not required for equilibrium of the roof. The roof does not have any loads applied. As Q_6 does not exist, Q_5 is not required for equilibrium of the sidewall. The sidewall does not have any applied loads. Therefore the floor (or chassis frame under the floor) has to carry the torsion moment from the front to the rear.

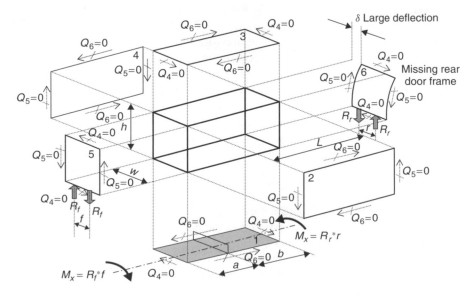

Figure 4.8 Box van with missing SSS in torsion.

It has already been stated in this chapter that loads applied normal to a surface cannot be effectively carried. The floor panel shown shaded in Figure 4.8 will not be very effective at carrying the torsion moment M_x. The condition described in Figure 4.1 of a moment about an axis in the SSS occurs which is not permissible. The floor must therefore be reinforced with a chassis frame that will be much more flexible than the box structure. This illustrates the advantage of an integral structure.

4.5   Examples of integral car bodies with typical SSS idealizations

A typical saloon (sedan) car body as shown in Figure 4.9 can be represented with SSSs as shown as a half-model in Figure 4.10. The half-model is shown for clarity and represents the right-hand side of the vehicle.

Surface (1) represents the floor cross-beam under the front seats and carries the seat loads together with the central longitudinal tunnel (2) (see Appendix). The floor panel itself (3) will not carry vertical loads but is essential for carrying shear in the torsion load condition. Under the rear seats, the cross-beam is represented by the surface (4) providing a similar function to beam (1). The luggage compartment floor, surface (5), is supported by the longitudinal (6) that carries both the luggage loads and the reaction R_R from the rear suspension. Usually the luggage compartment floor is at a higher level than the central floor panel and this is faithfully represented in the SSS

Figure 4.9 Saloon (sedan) car structure.

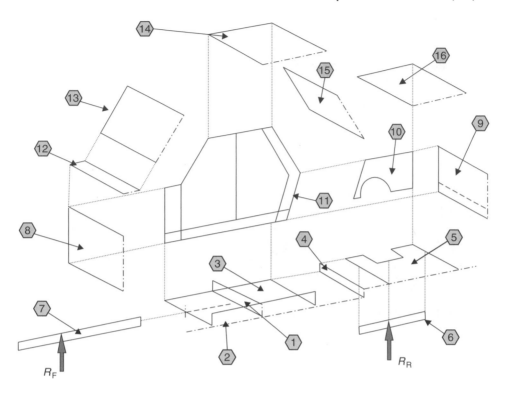

Figure 4.10 SSS half-model of sedan (saloon) structure.

model. At the front end the engine rail (7) carries the engine/transmission loads and the reaction R_F from the front suspension. The support for surface (7) is provided by the dash panel (8) and the floor cross-beam (1).

Moving up above the floor level, the dash panel (8) will be required to carry loads across to the sideframe (11) near to the 'A'-pillar. It will also form part of the torsion box (as demonstrated in Figure 4.7) for the torsion load case.

The rear panel (9) is required to react the load from the longitudinal (6) and to carry the loads out to the rear quarter panel (10). This in turn is attached to the main sideframe (11). Note, surfaces (10) and (11) could be regarded as one component.

The upper structure, consisting of upper dash or cowl plenum (12), the windscreen frame (13), the roof (14), the backlight frame (15) and the boot or trunk top frame (16), all form part of the torsion box in a similar way to the box van shown in Figure 4.7.

An important design feature of this model is the detail of surfaces (9) and (16), the rear panel and the top frame of the boot. Both of these surfaces are similar to that shown in Figure 4.4(c). The torsional stiffness of the body can be severely reduced if these structures are not designed with sufficient care.

Care is required to ensure adequate shear stiffness is obtained by raising the level of the boot 'lift-over' or by increasing the width of the sides. However, this needs to be balanced with consideration for a low 'lift-over' and wide opening dimension for the customer.

The SSS model shown in Figure 4.12 is a representation of the estate car or station wagon shown in Figure 4.11. Again the SSS model only shows the right-hand side of

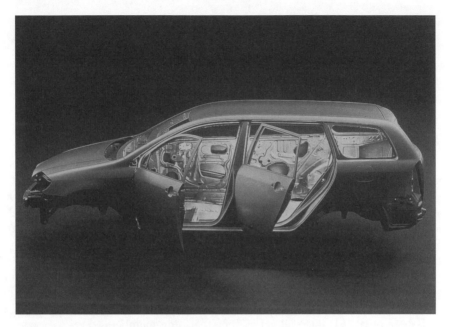

Figure 4.11 Corolla Fielder structure (courtesy of Toyota Motor Corporation).

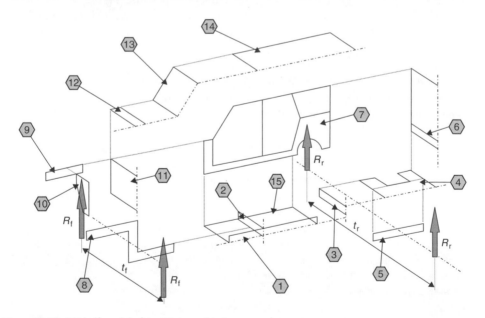

Figure 4.12 SSS half-model of an estate car (station wagon).

the vehicle for clarity. The floor of this vehicle has a central tunnel and a cross-beam under the front seats, similar to the saloon (sedan) shown in Figure 4.10. The rear seat cross-beam (3) is a substantial member leading to the raised rear floor (4).

Underneath the rear floor or luggage area, longitudinal members (5) run each side of the spare wheel-well connected to the rear seat cross-beam (3) and to the sill of the rear frame (6). In this vehicle the rear suspension mounting appears to be attached to the top of the rear wheel arch and close to the sideframe (7). Therefore it is assumed that the rear suspension reaction loads R_R are applied directly to the sideframe.

At the front there are substantial longitudinal members for the engine rails and at the top of the wings (fenders) which are represented by the SSSs (8) and (9). The suspension strut towers are attached at the top to the wing (fender) longitudinals and at the bottom to the engine rails. SSS (10) represents the strut tower and is effectively a cross-beam between the engine rail and the wing (fender) longitudinal. This type of SSS could also have been used for the rear suspension/wheel arch mounting. There are, of course, different ways of modelling a structure and the designer must use his/her judgement in selecting the most suitable SSSs. The engine rails (8) will be supported by the dash panel (11) and the cross-beam under the front seats (2). Note that these engine rails are not straight beams but are still plane structures and can be represented by SSSs. The upper wing rail (9) acts as a cantilever extending forward from the sideframe at the 'A'-pillar.

The upper members of the structure are the upper dash (12), the windscreen frame (13) and the roof panel (14). Only one other SSS remains to be noted, that is the floor (15). When this model is loaded with a torsion moment the SSSs numbered (4), (6), (7), (11), (12), (13), (14), (15) all act as shear carrying members.

Figure 4.13 shows a typical medium size delivery van while Figure 4.14 shows a possible SSS model. In Figure 4.14 the model represents a van with major longitudinals similar to a chassis frame. In this case the main bending loads are taken by the longitudinals (1) and (2), the cross-beams (3) and (4) and the sidewalls (5) and (6). If the cross-beams are relatively stiff compared to the longitudinals the sidewalls will carry the main longitudinal bending moment. The sidewalls are very deep compared with the chassis longitudinals and therefore much more stiff in bending. Chassis frames (longitudinals) are generally quite flexible in torsion, therefore the cross-beams (3)

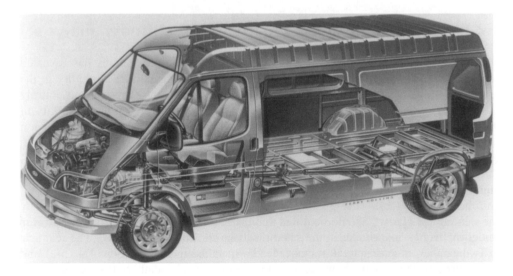

Figure 4.13 Van structure (courtesy of Ford Motor Co.).

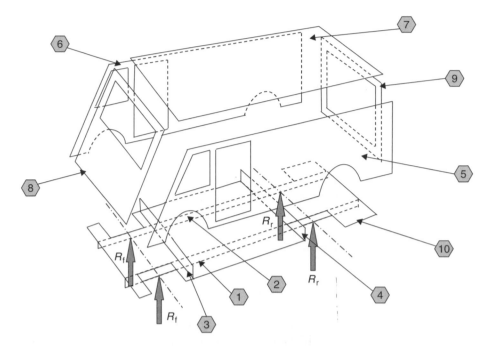

Figure 4.14 Simple structural surfaces representing a van structure with chassis frame.

and (4) will carry the torsion moments out into the sidewalls (5) and (6) of the van. The equilibrium of the sidewalls will be maintained by edge loads from the roof (7), the front panel (8), the rear door frame (9) and the floor (10). This loading will be similar to that shown in Figure 4.7 except that the moment is applied to the sidewalls rather than the end panels.

4.6   Role of SSS method in load-path/stiffness analysis

Modelling vehicles with the SSS method as shown in these examples has revealed problem areas in the design concepts. In Figure 4.8 it was noted that flexibility in the rear door frame of a simple box results in the torsion moment being carried entirely in the floor or chassis frame. In the saloon car model of Figure 4.10 the flexibility in the rear panel due to the lowering of the boot sill and the open end of the boot top frame were noted. The model of the estate car in Figure 4.12, when loaded in torsion, relies on an adequate stiff rear door frame in order to transfer shear loads into the roof. Similarly, the windscreen frame must have adequate stiffness. The windscreen frame is composed of the windscreen pillar or 'A'-pillar (side), the front roof header rail (top), and the cowl (bottom). This assembly acts as a ring frame and is further stiffened by the windscreen glass. It should be noted that if the surrounding frame has low stiffness the glass may be loaded excessively and result in cracking of the glass.

Two further structural weaknesses are shown in the models of Figure 4.15(a) and (b). These reveal problems in bending when the vehicle is loaded. At (a) the lack of an SSS in the horizontal plane at the upper dash panel will result in normal loading to the

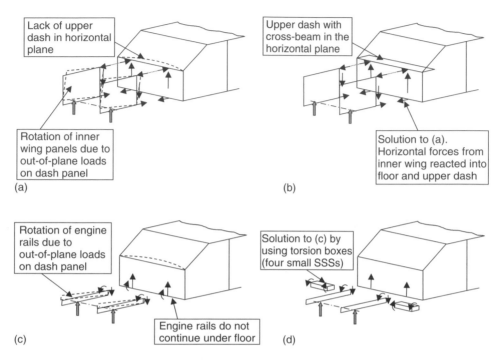

Lack of upper dash in horizontal plane

Rotation of inner wing panels due to out-of-plane loads on dash panel

(a)

Upper dash with cross-beam in the horizontal plane

Solution to (a). Horizontal forces from inner wing reacted into floor and upper dash

(b)

Rotation of engine rails due to out-of-plane loads on dash panel

Engine rails do not continue under floor

(c)

Solution to (c) by using torsion boxes (four small SSSs)

(d)

Figure 4.15 Examples of missing SSSs and solutions.

dash panel resulting in large deflections as the panel is very flexible in this direction according to the principles of SSS. A solution to this problem is to include an upper dash panel that has good stiffness in the horizontal plane as shown at (b). The lower edge forces from the inner wing panel can be reacted in the floor. At (c) the engine rail is not continued under the centre floor panel to be attached to the cross-beam under the front seats. An alternative method of supporting the engine rail is shown at (d) where a torsion box (or 'torque box') connects the engine rail to the sideframe. The bending moment in the engine rails is carried as shear in the surfaces of the torsion boxes. Each torsion box can be modeled as four SSSs. The vertical force is carried into the dash panel.

The above examples illustrate how the SSS method can be used to identify possible weaknesses in a structural concept. By carefully evaluating the loads necessary to maintain the equilibrium of each SSS and the action/reaction forces acting between them, structural integrity can be enhanced.

However, no theoretical method can provide a full understanding of the structural load paths and the SSS method is no exception. The method yields an exact solution if the structure and its component parts are statically determinate but passenger cars are highly redundant structures. Therefore in practice it is necessary to make simplifying assumptions. This can be illustrated by considering the rear of a saloon car as shown in Figure 4.16 where the torsion load case is applied. In order to carry the torsion moment it is necessary to transfer shear loads from the floor to the roof and Figure 4.16(a) shows two possible load paths (i.e. one is redundant). In Figure 4.16(b) the shear load path is via the panel behind the rear seats, the rear parcel shelf and the backlight.

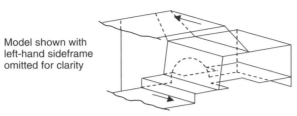

Model shown with
left-hand sideframe
omitted for clarity

(a) Saloon (sedan) rear structure with two shear paths (via rear seat panel, parcel tray, backlight and via boot floor, rear panel, boot (trunk) top frame, backlight)

Arrows indicate the shear loads on the cross-members/panels

(b) Shear transferred via the rear seat panel, rear parcel tray, backlight

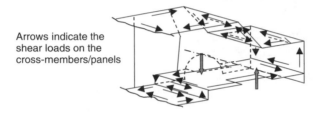

Arrows indicate the shear loads on the cross-members/panels

(c) Shear transferred via the boot floor, rear panel and boot top frame, backlight

Figure 4.16 Saloon (sedan) car rear structure.

Figure 4.16(c) illustrates an alternative load path via the boot floor, the rear panel, the top frame of the boot (trunk) and the backlight. The SSS method can only be solved by assuming one of these load paths. In practice the loads will be shared between the two load paths, the proportions carried by each will depend on the relative stiffness. The relative stiffness will be influenced by the particular load path strategy selected by the body structure engineer and constrained by packaging space, styling, manufacturing, and other considerations.

In spite of this limitation, the SSS method has shown the type of loading that occurs on each structural element. It also helps illustrate the implications of one load path versus another. In this example it illustrates that all the components must be designed to carry shear forces, and with this understanding the designer is encouraged to focus on achieving the design features that will provide adequate shear strength and shear stiffness.

Appendix to Chapter 4

Edge load distribution for a floor with a simple grillage

The distribution of the payload F_{pl} to the edge loads K_1, K_2 and K_3 (see Figure 4.A1) can be determined as follows:

From Figure 4.A1(a)

Let F_{plc} be the portion of payload carried by the cross-beam. The cross-beam is loaded at mid-span therefore from standard beam formulae the deflection of the cross-beam at the load point is:

$$\delta_{cc} = \frac{F_{plc} w^3}{48 E I_c} = \frac{2 K_1 w^3}{48 E I_c}$$

From Figure 4.A1(b) and (c) we can apply the unit load method to determine the deflection of the longitudinal beam at the load application point.

For the applied loads we obtain equations for the bending moments.

Between D and C   **Between E and C**

$M = K_2 x$ $M = K_3 x$

For the unit load applied at the point C, the equations for the bending moments are:

$$m = \frac{b}{(a+b)} x \qquad\qquad m = \frac{a}{(a+b)} x$$

From the unit load method the deflection at C for the longitudinal beam is given by:

$$\delta_{cl} = \sum \int \frac{M m}{E I} dx$$

$$\delta_{cl} = \frac{1}{E I_1} \left[\int_0^a \frac{b}{(a+b)} K_2 x^2 \, dx + \int_0^b \frac{a}{(a+b)} K_3 x^2 \, dx \right]$$

$$= \frac{ab}{3 E I_1 (a+b)} [a^2 K_2 + b^2 K_3]$$

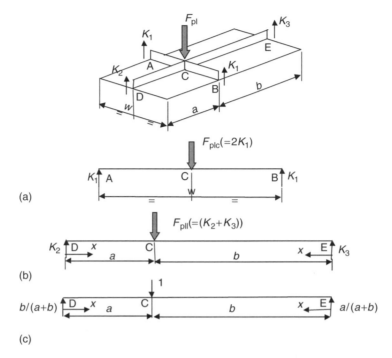

(a)

(b)

(c)

Figure 4.A1 Floor with longitudinal and cross-beams.

The deflection of the cross-beam and the longitudinal at the joint must be equal.

$$\therefore \quad \delta_{cc} = \delta_{cl}$$

$$\frac{2K_1 w^3}{48EI_c} = \frac{ab}{3EI_1(a+b)}[a^2 K_2 + b^2 K_3]$$

With similar materials for both beams:

$$\frac{K_1 w^3}{8I_c} = \frac{ab}{I_1(a+b)}[a^2 K_2 + b^2 K_3] \tag{4.A1}$$

For the longitudinal beam:

$$K_2 a = K_3 b \tag{4.A2}$$

For the two beams together:

$$2K_1 + K_2 + K_3 = F_{pl} \tag{4.A3}$$

These three equations can be solved to obtain the three edge loads K_1, K_2 and K_3.
    For example:

$$K_2 = \frac{F_{pl} w^3 b I_1}{[I_1(a+b)w^3 + 16I_c a^2 b^2]}$$

If the floor is square and the beams are placed at the longitudinal centreline and at the mid-side position

i.e. $\qquad\qquad\qquad\qquad\qquad a = b \quad \text{and} \quad w = 2a$

Substituting these values in (4.A1):

$$\frac{K_1 (2a)^3}{8I_c} = \frac{a^4}{I_1 (2a)} [K_2 + K_3]$$

$$\frac{K_1}{I_c} = \frac{K_2 + K_3}{2I_1}$$

or $\qquad\qquad\qquad \dfrac{2K_1}{(K_2 + K_3)} = \dfrac{I_c}{I_1}$

In this case the payload F_{pl} is shared between the two beams in proportion to the beam second moments of area. Provided the floor does not depart too much from the square condition this ratio can be used for an approximate load distribution.

5

Standard sedan (saloon) – baseline load paths

Objectives

- To demonstrate that a car structure can be represented by simple structural surfaces (SSS).
- To introduce the load paths in a sedan structure for different load cases.

5.1   Introduction

The structures of different sedan passenger car structures vary according to size, vehicle layout ('package') and type and to the particular design and assembly methods of different manufacturers. Nevertheless, the nature of integral construction dictates that there will also be similarities.

In this chapter, a simplified car structure referred to as the 'standard sedan' is used as the basis for an introductory discussion of load paths in integral car structures. For further clarity, simplification of the payload and of the suspension input loads is also made.

The major load cases (especially bending and pure torsion) are described separately because they each make very different structural demands on the individual simple structural surfaces in the vehicle body. Actual road loads will be a combination of these cases, and the calculation results can be obtained as appropriate.

Since the main interest in this chapter is in load paths through the body structure, the evaluation of stresses, and the sizing of components is deferred to Chapters 7, 11 and 12.

5.1.1   The standard sedan

The standard sedan, Figure 5.1, consists of a 'closed box' passenger compartment, comprising floor, roof, sideframes, front and rear bulkheads and windscreen. For simplicity, all of these surfaces are assumed to be plane.

The suspension loads, at both front and rear, are carried on deep, stiff boom/panel cantilevers attached to the ends of the compartment. These are a simplified representation

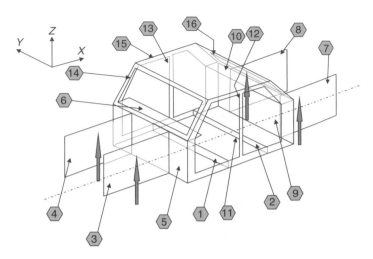

Figure 5.1 Baseline model.

of the inner wing panels. The booms represent the lower rails (e.g. for engine mounting) and the upper flanges. To keep the model simple, the suspension tower loads are fed directly into the webs of the cantilevers.

Where it is necessary to carry out-of-plane loads, 'supplementary SSSs' (acting as beams) are provided, for example as floor cross-members and parcel shelves. To reduce complexity in the standard sedan, the parcel shelves are not considered to be part of the surface of the passenger compartment 'torsion box' for the torsion load case. Both parcel trays are, however, necessary for carrying loads in the bending load case, and for carrying loads into the 'torsion box' in the torsion case.

A more detailed discussion of different and more realistic structural features, including alternative end structure assemblies, is reserved for Chapter 6.

The 16 simple structural surfaces in the standard sedan, Figure 5.1, are as follows:

1. Transverse floor beam (front) carrying the front passengers.
2. Transverse floor beam (rear) carrying the rear passengers.
3. and 4. Inner wing panels carrying the power-train and supported by the front suspension.
5. Dash panel–transverse panel between passengers and engine compartment.
6. Front parcel shelf.
7. and 8. Rear quarter panels carrying luggage loads and supported by the rear suspension.
9. Panel behind the rear seats.
10. Rear parcel shelf.
11. Floor panel
12. and 13. Left-hand and right-hand sideframes.
14. Windscreen frame.
15. Roof panel.
16. Backlight (rear window) frame.

These SSSs will be shown to be sufficient to carry the two fundamental load cases of bending and torsion. Some additional SSSs will be necessary for other load cases as described in section 5.4. Alternative SSSs may be necessary when modelling particular vehicles as discussed in Chapter 8. The vehicle engineer must use his knowledge and make his/her own subjective assessment for the model that best represents a particular structure.

5.2   Bending load case for the standard sedan (saloon)

5.2.1   Significance of the bending load case

Figure 5.2 shows the baseline loads that are considered for the bending case. The main loads of power-train F_{pt}, the front passengers/seats F_{pf}, the rear passengers/seats F_{pr}, and the luggage F_ℓ only are considered. The magnitude of the loads is the weight of the component factored by a dynamic load factor as described in Chapter 2. It should be noted that all these loads are applied in the planes of simple structural surfaces. It is essential this condition is achieved in order to ensure sufficient strength and stiffness can be provided through the structure. The bending and shear loads on each component can be determined and from these satisfactory stress levels can be determined.

5.2.2   Payload distribution

The passenger car structure when viewed in side elevation (Figure 5.3) can be considered as a simply supported beam, the supports are at the front and rear axles. First, the masses of these components and their longitudinal and lateral positions in the vehicle must be known, then the front and rear suspension reaction forces can be obtained. Referring to Figure 5.3, by taking moments about the rear suspension mounting the front suspension reaction is:

$$R_F = \frac{F_{pt}(L + l_{pt}) + F_{pf}(L - l_{pf}) + F_{pr}(L - l_{pr}) - F_\ell l_\ell}{L} \qquad (5.1)$$

where l_{pt}, l_{pf}, l_{pr} and l_ℓ are defined in Figure 5.3.

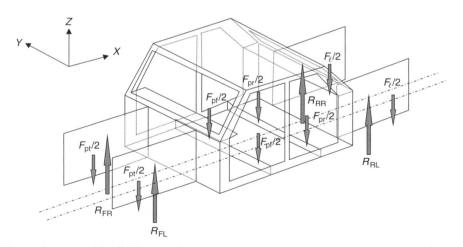

Figure 5.2 Baseline model – bending loads.

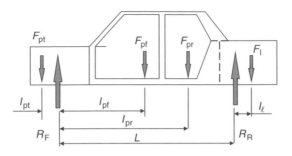

Figure 5.3 Payload distribution.

Similarly, taking moments about the front suspension mounting:

$$R_R = \frac{F_{pf}l_{pf} + F_{pr}l_{pr} + F_\ell(L + l_\ell) - F_{pt}l_{pt}}{L} \tag{5.2}$$

Check the answers by verifying $R_R + R_F = F_{pt} + F_{pf} + F_{pr} + F_\ell$.

When investigating a particular vehicle as described in Chapter 8 many more components can be considered making a more accurate model. Typical items that may be included are: front bumper, radiator, battery, instrument panel/steering column, exhaust, fuel tank, spare wheel, rear bumper and distributed loads due to the weight of the body structure. If these are to be included, the positions of all these components as well as their masses must be known. The suspension reactions can be calculated with a similar procedure.

5.2.3   Free body diagrams for the SSSs

Developing the model of Figure 5.2 into the 'exploded' view shown in Figure 5.4, it can be seen that we require edge loads and end loads to ensure all SSSs are in equilibrium. These edge/end loads are indicated by the forces P_1 to P_{13}.

When beginning the bending analysis it is essential to start at the central floor area. In this simple model it is assumed that the passenger loads are carried by the two transverse floor beams SSS (1) and (2). (See the Appendix in Chapter 4 for an alternative floor model.) These floor beams are supported at each end by the side-frame forces P_1 and P_2. Note that there is an equal but opposite force acting on the sideframe.

Consider now the inner front wings SSS (3) and (4), the loads acting on these are the loads from the power-train $P_t/2$ and from the front suspension R_{FL}. The applied loads $F_{pt}/2$ and R_{FL} are held in equilibrium by the end loads P_4 and P_5 and by the edge (shear) load P_3. The shear load P_3 reacts into the dash panel while the end load P_4 reacts into the front parcel shelf (6), and P_5 into the floor panel (11). These forces can be obtained by the equations of statics, i.e. resolving forces and taking moments.

When building the SSS model representing the vehicle it must always be remembered that the forces must act in the plane of an SSS. Failure to provide sufficient SSSs to satisfy this requirement soon reveals a weakness or unsatisfactory load path in the structure. Note, the horizontal SSS (6) is necessary in order to carry force P_4.

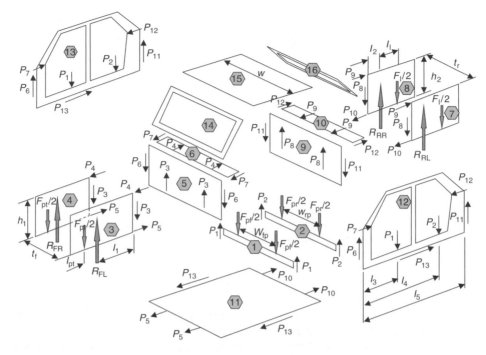

Figure 5.4 Baseline model – bending case, end and edge loads.

By working through the individual SSSs of this model shown in the next section it will be realized that there are sufficient forces to achieve equilibrium for each component and that all loads act in the planes of the SSSs.

5.2.4   Free body diagrams and equilibrium equations for each SSS

Transverse floor beam (front) (1)
Resolving forces vertically and by symmetry (loads are assumed to be applied symmetrically about the vehicle longitudinal centreline):

$$P_1 = F_{pf}/2 \tag{5.3}$$

Transverse floor beam (rear) (2)
Resolving forces vertically and by symmetry:

$$P_2 = F_{pr}/2 \tag{5.4}$$

Left and right front inner wing panel (3) and (4)
Resolving forces vertically for the left-hand panel:

$$P_3 = R_{FL} - F_{pt}/2 \tag{5.5}$$

A similar equation is obtained for the right-hand panel.

Taking moments about the rear lower corner:

$$P_4 = \{R_{FL}l_1 - F_{pt}(l_1 + l_{pt})/2\}/h_1 \qquad (5.6)$$

Resolving forces horizontally:

$$P_5 = P_4 \qquad (5.7)$$

Dash panel (5)
Equal and opposite reaction forces P_3 to those on the wing panels act on this SSS. Resolving forces vertically and by symmetry:

$$P_6 = P_3 \qquad (5.8)$$

Front parcel shelf (6)
Equal and opposite reaction forces P_4 to those on the inner wing panels act on this SSS. Resolving forces horizontally and by symmetry:

$$P_7 = P_4 \qquad (5.9)$$

Rear quarter panels (7) and (8)
Resolving vertically for the left-hand panel:

$$P_8 = R_{RL} - F_\ell/2 \qquad (5.10)$$

Taking moments about the front lower corner:

$$P_9 = \{R_{RL}l_2 - F_\ell(l_1 + l_2)/2\}h \qquad (5.11)$$

Resolving forces horizontally:

$$P_{10} = P_9 \qquad (5.12)$$

Similar equations apply for the right-hand panel.

Panel behind the rear seats (9)
Resolving forces vertically and by symmetry:

$$P_{11} = P_8 \qquad (5.13)$$

Rear parcel shelf (10)
Resolving forces horizontally and by symmetry:

$$P_{12} = P_9 \qquad (5.14)$$

Floor panel (11)
Reaction forces P_5 from the inner front wing panels and forces P_{10} from the rear quarter panel are applied to this SSS. These will not necessarily be equal so additional forces P_{13} are required acting at the sides which react on the sideframes. It will be assumed these forces act in the direction shown in Figure 5.4 although when numerically evaluated, these may be negative (i.e. in the opposite directions). Resolving forces horizontally:

$$2P_{13} = 2(P_{10} - P_5) \qquad (5.15)$$

Left-hand and right-hand sideframes (12) and (13)

Both sideframes are loaded identically. Examining the forces acting on the sideframes shows that these have already been obtained from equations (5.3), (5.4), (5.8), (5.9), (5.13), (5.14) and (5.15). However, it is necessary to check that equilibrium conditions are satisfied by applying the equations of statics. Experience has shown that it is essential to make this equilibrium check as errors do occur with the use of the many equations.

Resolving forces vertically:

$$P_6 - P_1 - P_2 + P_{11} = 0 \tag{5.16}$$

Resolving forces horizontally:

$$P_7 + P_{13} - P_{12} = 0 \tag{5.17}$$

Moments may be taken about any point but in order to reduce the algebra it is better to take moments about a point where two forces act. For example, take moments about the lower corner of the windscreen pillar where P_6 and P_7 act. This simplifies the equation by eliminating two terms.

Moments about the lower corner of the windscreen:

$$P_1 l_3 + P_2 l_4 - P_{11} l_5 - P_{12}(h_2 - h_1) = 0 \tag{5.18}$$

In practice some rounding errors due to difficulties in defining the exact positions of each force may occur.

It should now be noted that windscreen frame (14), roof panel (15), and back-light (16) SSSs are not subject to any load for this bending case.

5.2.5   Shear force and bending moment diagrams in major components – design implications

Now that the forces on each SSS have been obtained the shear force and bending moment diagrams can be drawn.

Figure 5.5(a) shows the loading, shear force and bending moment diagrams for the front transverse floor beam. The rear transverse floor beam although not shown is loaded in a similar manner. It should be realized that the beam is simply supported at its ends where it is attached to the sill member. The justification for this is given in Chapter 11. Design of this joint must be suitable for carrying the vertical shear force P_1. The centre section has a constant bending moment and must be designed to provide suitable bending properties. Note that the positive bending moment means the beam is subject to a sagging moment.

In Figure 5.5(b) the loading on the dash panel is shown, a similar condition applies to the panel behind the rear seats. In contrast to the floor beams this panel is subject to a negative bending moment or hogging moment. The panel behind the rear seats is also loaded in this manner. These panels are relatively deep so bending stresses and deflections will be small although stiffening at the top and bottom edges will be necessary to prevent buckling. The outer sections carrying the shear force will probably require swaging to prevent shear buckling.

The loading conditions on the front and rear parcel shelves are once again similar. The front shelf shown in Figure 5.5(c) is loaded such that in plan view it is deflected

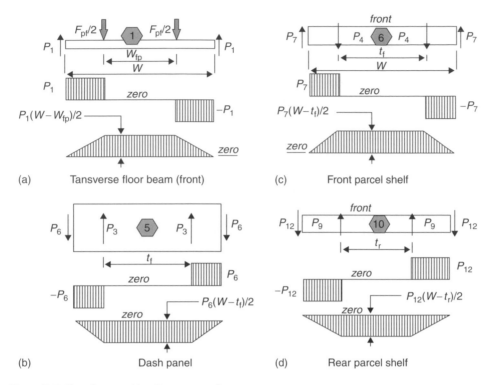

Figure 5.5 Shear force and bending moment diagrams.

towards the rear (+ve bending moment) while the rear shelf in Figure 5.5(d) is deflected forward (−ve bending moment). Both these shelves must have good bending properties in the centre and adequate shear connections to the sideframe.

Considering the loads at the front and rear of the structure we have the conditions shown in Figure 5.6. At (a) the loading on the front inner wing panel is shown and at (b) that on the rear quarter panel. With modern transverse engined cars the engine centre of gravity is forward of the front suspension so there is a hogging moment forward of the suspension but a sagging moment where the wing panel is attached to the dash panel etc. A similar situation occurs at the rear of the vehicle where the luggage is placed behind the rear suspension mountings causing a hogging moment. This may well change to a sagging moment at the attachment of the rear quarter panel to the panel behind the rear seats, depending on magnitudes of forces and dimensions. Both front inner wing panels and rear quarter panels must be designed to carry the indicated shear forces. The horizontal reaction forces P_4, P_5, P_9 and P_{10} for these SSSs must be distributed along the top and bottom edges by means of stiffeners as indicated by the dashed lines. Buckling of these deep members due to shear forces can be prevented by suitable stiffening swages.

The floor panel loading shown in Figure 5.7(a) indicates that the outer sides are subject to shear forces. The shear force P_{13} is applied over a long length (l_5 shown in Figure 5.4) hence the shear stresses are usually small. Local panel stiffening will be necessary for other reasons (e.g. to prevent panel vibrations and resist normal loads).

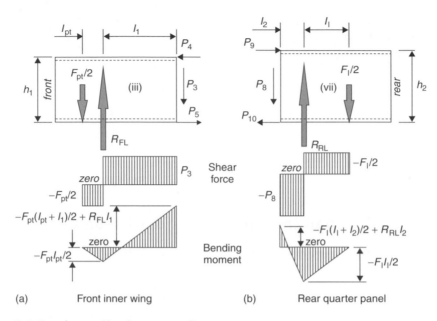

(a)   Front inner wing   (b)   Rear quarter panel

Figure 5.6 Shear force and bending moment diagrams.

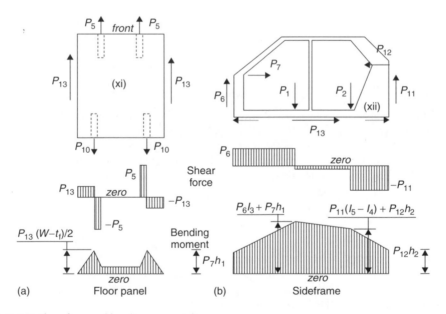

(a)   Floor panel   (b)   Sideframe

Figure 5.7 Shear force and bending moment diagrams.

Although the diagram indicates bending moments these are not of significance as l_5 is large.

The end loads P_5 and P_{10} in practice cannot be applied as point loads to the front and rear edges of the floor. The stiffener necessary at the base of the front inner wing panel and the rear quarter panel will need to be extended along the floor panel as shown with dashed lines (Figure 5.7(a)) (see also Chapter 11). The sideframe, Figure 5.7(b), is the main structural member providing bending strength and stiffness. The internal load distribution through the elements of the sideframe is not considered here (but see Chapter 7). From equation (5.15) the edge load P_{13} is the difference between P_{10} and P_5 which in turn can be shown equal to the difference between P_{12} and P_7 using equations (5.12)/(5.14) and (5.7)/(5.9). Therefore the front and rear ends of the sideframe may be considered to be subject to bending moments $P_7 h_1$ and $P_{12} h_2$, respectively. The centre part of the sideframe is also subject to increased bending due to the shear forces P_6 and P_{11}. Therefore, the loading on the cantrail will be a combination of bending and compression and the loading on the sill a combination of bending and tension. Note also additional local bending will occur on the sill from the loads P_1 and P_2.

5.3   Torsion load case for the standard sedan

5.3.1   The pure torsion load case and its significance

On the road, the car is subjected to torsion when a wheel on one side strikes a bump or a pot-hole, causing different wheel reactions on each side of the axle. This is the vertical asymmetric load case (see Chapter 2) which gives a combination of bending and torsion on the vehicle.

For calculation purposes, the torsion component of the asymmetric vertical case is considered *in isolation*, as the *pure torsion load case*. Equal and opposite loads R_{FT} are applied to the front left and right suspension towers, thus causing a couple T about the vehicle centreline. This is reacted by an equal and opposite couple at the rear suspension points so that the vehicle is in pure torsion (see Figure 5.8). The SSS edge loads Q, resulting from this, are then calculated.

Clearly, the pure torsion load could not be experienced on the road, since there cannot be a negative wheel reaction. However, if road case loads are required, then the SSS edge loads Q from the pure torsion case could be combined with the edge loads P from the bending load case by suitable factoring and addition (see Chapter 2).

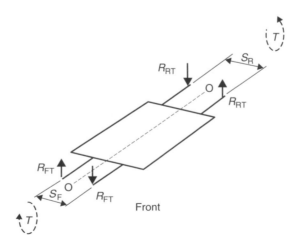

Figure 5.8 Vehicle in pure torsion.

The significance of the pure torsion load case is that it applies edge forces on the individual simple structural surfaces that are completely different from those experienced in the bending case. The torsion stiffness (and hence also the torsional fundamental vibration frequency) of a vehicle body is often used as a benchmark of its structural competence.

The torsion case is found to be a stringent one. For torsion, the keys to a weight efficient integral sedan structure are:

1. a closed ('boxed') system of simple structural surfaces, in shear, in the passenger compartment, and
2. as in other load cases, continuity of the load paths at the dash, where the suspension loads are fed from the end structures into this 'torsion box'.

In this section, the baseline *standard sedan* with closed torsion box is discussed first. Later, the *'faux sedan'*, with at least one missing simple structural surface in the passenger compartment is considered. The missing surface(s) can have a disastrous effect on the torsion performance of the body. Remedies to the faux sedan's deficiencies are suggested. Different and more realistic suspension mounting structures, on the ends of the body, are discussed in Chapter 6.

5.3.2   Overall equilibrium of vehicle in torsion

The torque T is applied about axis O−O as couple $R_{FT}S_F$ at the front suspension. This must be balanced by an equal and opposite couple $R_{RT}S_R$ due to the reaction forces R_{RT} at the rear suspension. See Figure 5.8:

$$T = R_{FT}S_F = R_{RT}S_R$$

Hence

$$R_{FT} = T/S_F \qquad R_{RT} = R_{FT}\ S_F/S_R = T/S_R$$

5.3.3   End structures

(a) Front and rear inner fenders
On the right-hand fender as shown in Figure 5.9, the suspension load acts upward. This is reacted by an equal downward force on the panel where it is joined to the bulkhead.

For moment equilibrium, the couple caused by the offset L_1 of forces R_{FT} is balanced by complementary shear forces P_{FT} at top and bottom of the panel:

$$R_{FT}L_1 = P_{FT}h_1 \quad \text{thus} \quad P_{FT} = R_{FT}L_1/h_1 = TL_1/(S_Fh_1)$$

Forces P_{FT} are reacted by equal forces, fed into the top and bottom flanges as shear flows (see later). These will in turn be reacted by axial forces P_{FT} in the flanges, as shown, where the flanges meet the passenger compartment.

The upper flange is of thin sheet material, so that the reaction P_{FT} will be concentrated at the junction between the web and the flange so that the latter can be treated as a 'boom' (see Chapter 4). The lower flange is usually the 'engine mounting rail' consisting of a substantial box member. This may also be treated as a boom here.

The left-hand fender will behave similarly, but with the forces in opposite directions.

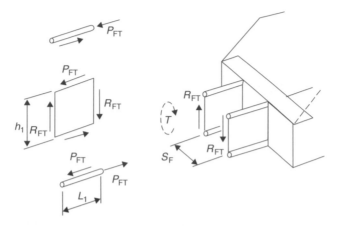

Figure 5.9 Frontal structure.

In the standard sedan, the rear inner fenders will behave in an identical way to the front ones, with the forces in the appropriate directions (see Figure 5.12). The vertical shear force on the rear fender is R_{RT} and the reaction forces in the top and bottom flanges P_{RT} are:

$$P_{\mathrm{RT}} = R_{\mathrm{RT}}L_2/h_2 = T L_2/(S_R h_2)$$

where L_2 = loaded length and h_2 = height of rear inner fender.

(b) Dash

Torque T is applied to the engine bulkhead by the reactions R_{FT} from the fender webs acting at a separation S_F giving a couple $P_{\mathrm{FT}}S_F = T$ as shown in Figure 5.10.

The reactions P_{FT} to the fender upper flange forces are carried by the parcel shelf which must be stiff in the appropriate direction. The reactions to the engine rail forces P_{FT} pass to the floor as in-plane forces in the directions shown in Figure 5.12.

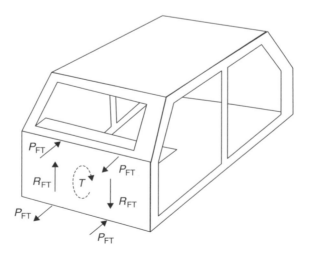

Figure 5.10 Forces on dash panel.

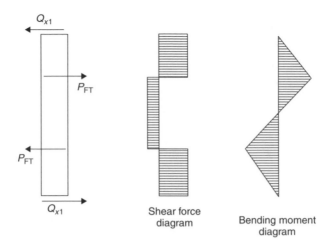

Figure 5.11 Parcel shelf.

The rear fender forces are reacted in an identical way. Owing to the directions of the suspension forces in the torsion case, forces P_{RT} apply a couple to the parcel shelves of $P_{RT}S_R$. This is also the case for the floor.

(c) Parcel shelf/upper dash
The parcel shelf acts as a beam, carrying the couple $P_{FT}S_F$ out to the sideframe at the mid A-pillars.

The end forces Q_{X1} form a couple to balance this couple, thus:

$$Q_{X1}B = P_{FT}S_F \quad \text{giving} \quad Q_{X1} = P_{FT}S_F/B \qquad (5.19)$$

The shear forces in the parcel shelf (acting in a horizontal plane) and the bending moments (acting about a vertical axis) are as shown in Figure 5.11. By similar reasoning the rear parcel shelf reaction forces Q_{X2} are:

$$Q_{X2} = P_{RT}S_R/B \qquad (5.20)$$

5.3.4   Passenger compartment

A general view of the SSS edge forces in the passenger compartment is given in Figure 5.12.

(a) Engine bulkhead
The shear forces R_{FT} from the fender webs are reacted on the engine bulkhead. This creates a couple $T = R_{FT}S_F$

This couple is reacted by shear forces Q_1 and Q_2 acting on the edges of the bulkhead, forming couples Q_1h_1 and Q_2B.

For lateral force equilibrium: $Q_{1 \text{ TOP}} = Q_{1 \text{ BOTTOM}} = Q_1$

For vertical force equilibrium: $Q_{2 \text{ LEFT}} = Q_{2 \text{ RIGHT}} = Q_2$

For moment equilibrium: $T = Q_1h_1 + Q_2B$ $\qquad (5.21)$

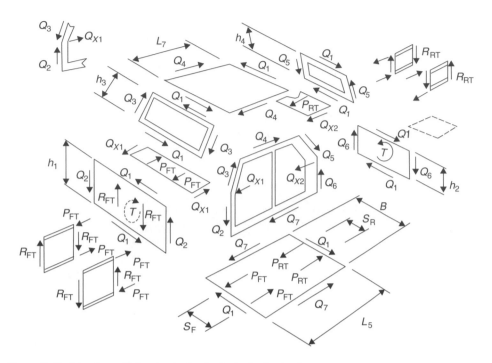

Figure 5.12 Edge forces in the standard sedan in the torsion load case.

(b) Front windshield

Shear force Q_1 from the top of the engine bulkhead is reacted by an equal force on the bottom of the windshield frame. For horizontal force equilibrium, this will be balanced by an equal force Q_1 on the top edge of the frame.

These forces Q_1 form a couple $Q_1 h_3$ which must be balanced by a couple $Q_3 B$ from complementary shear forces Q_3 on the sides of the frame.

$$Q_1 h_3 - Q_3 B = 0 \tag{5.22}$$

This frame achieves its shear stiffness from the local in-plane bending stiffness (i.e. about an axis *normal* to the plane of the windshield frame). It is thus working as a 'ring beam', see Chapter 4 and section 5.3.6.

(c) Roof

Shear force Q_1 is fed to the front of the roof from the top of the windshield frame. This will be balanced by edge force Q_1 at the rear. These forces form a couple $Q_1 L_7$ which must be balanced by a complementary couple $Q_4 B$ from the edge forces Q_4 on the roof sides. Thus, for moment equilibrium:

$$Q_1 L_7 - Q_4 B = 0 \tag{5.23}$$

(d) Backlight (rear window) frame

Force Q_1 from the rear of the roof is reacted by an equal and opposite force on the top of the rear window frame and this, in turn, is balanced by a force Q_1 on the bottom of the window frame. The couple $Q_1 h_4$ from these forces is balanced by the complementary couple $Q_5 B$ from edge forces Q_5 on the sides of the backlight frame.

$$\text{For moment equilibrium: } Q_1 h_4 - Q_5 B = 0 \tag{5.24}$$

(e) Rear seat bulkhead

Shear force Q_1 is passed from the backlight frame to the top of the rear bulkhead. The top and bottom edge forces Q_1 on the rear bulkhead and forces Q_6 on its sides form couples which are balanced by the couple $T = R_{RT} S_R$ applied by the rear fender webs to this panel.

$$\text{Thus for moment equilibrium: } T = Q_1 h_2 + Q_6 B \tag{5.25}$$

As before $Q_{1\ \text{TOP}}$ and $Q_{1\ \text{BOTTOM}}$ are equal and opposite, as are $Q_{2\ \text{LEFT}}$ and $Q_{2\ \text{RIGHT}}$. Because of the externally applied torque T, the edge forces Q_1, Q_6 do not form opposing couples (although *overall* the panel must be in moment equilibrium). As on all such surfaces with additional external forces or moments, force Q_1 is the *net* force on the top of this panel. The rear seat bulkhead is sometimes a continuous panel, or it may consist of a truss structure, formed by punching triangular holes leaving diagonal members forming a 'triangulated truss' SSS (Figure 5.13).

(f) Floor

This receives equal and opposite edge forces Q_1 from the front and rear bulkheads and Q_7 at the sides, from the sideframes. The forces Q_1 and Q_7 are oriented so as to form complementary couples $Q_1 L_5$ and $Q_7 B$, except that now additional forces P_{FT} and P_{RT} from the lower longitudinals of the front and rear fenders cause additional couples $P_{FT} S_F$ and $P_{RT} S_R$. All the pairs of forces Q_1, Q_7, P_{FT}, P_{RT} balance out for *force* equilibrium.

$$\text{For } moment \text{ equilibrium: } Q_1 L_5 - Q_7 B = P_{FT} S_F + P_{RT} S_R \tag{5.26}$$

Often, the engine rails run for some distance under the floor, and this gives a good connection path (in shear) for force P_{FT} between the engine rail and the floor.

Figure 5.13 Rear seat aperture with diagonal braces (courtesy Vauxhall Heritage Archive).

(g) Sideframes

The edges of the sideframe react edge forces $Q_2Q_3Q_4Q_5Q_6Q_7$ from the surfaces attached to it, and Q_{x1} and Q_{x2} from the front and rear parcel shelves, as seen in Figure 5.14. The sideframes on opposite sides of the vehicle experience identical edge loads, but in opposite directions. Force and moment equilibrium will be obeyed. Clearly, the sideframes are crucial in 'gathering' the edge forces from the other surfaces.

For moment equilibrium, take moments about an arbitrary point G distance X behind the lower A-pillar centreline and Z above the rocker centreline. If r_2 to r_7 are the moment arms of forces Q_2 to Q_7 about point G as in Figure 5.14, then:

$$-r_2Q_2 + r_3Q_3 + r_4Q_4 + r_5Q_5 - r_6Q_6 + r_7Q_7$$
$$= Q_{X1}(h_1 - Z) + Q_{X2}(h_2 - Z)\ldots \tag{5.27}$$

The moments of most of the forces about G are shown in Table 5.3(a) (taking clockwise moments as positive).

The forces Q_3 and Q_5 act on the upper A- and C-pillars, which are at angles θ_A and θ_C from the vertical. These forces can be resolved into vertical and horizontal components. Noting that the line of action of Q_5 passes through the C-pillar waist joint (point E in Figure 5.15) and that of Q_3 passes through the A-pillar waist joint (point B in Figure 5.15), then the moments of these forces are given in Table 5.3(b).

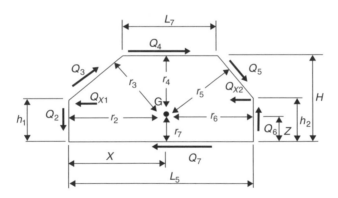

Figure 5.14 Sideframe.

Table 5.3(a)

Edge force	Moment about G	Moment arm about G
Lower A-pillar	$-Q_2X$	$r_2 = X$
Header (roof rail)	$Q_4(H - Z)$	$r_4 = H - Z$
Lower C-pillar	$-Q_6(L_5 - X)$	$r_6 = L_5 - X$
Rocker	Q_7Z	$r_7 = Z$

Parcel shelf reactions	Moment about G	
Front	$-Q_{x1}(h_1 - Z)$	$h_1 - Z$
Rear	$-Q_{x2}(h_2 - Z)$	$h_2 - Z$

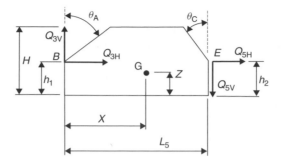

Figure 5.15 Components of forces Q_3 and Q_5.

Table 5.3(b)

Force (Figure 5.15)	Component	Moment about G	Moment arm about G
Q_{3H}	$Q_3 \sin(\theta_A)$	$(h_1 - Z)Q_3 \sin(\theta_A)$	$r_{3H} = (h_1 - Z)$
Q_{3V}	$Q_3 \cos(\theta_A)$	$(X)Q_3 \cos(\theta_A)$	$r_{3V} = (X)$
Q_{5H}	$Q_5 \sin(\theta_C)$	$(h_2 - Z)Q_5 \sin(\theta_C)$	$r_{5H} = (h_2 - Z)$
Q_{5V}	$Q_5 \cos(\theta_C)$	$(L_5 - X)Q_5 \cos(\theta_C)$	$r_{5V} = (L_5 - X)$

Force	Moment about G	Moment arm about G
Q_3	$Q_3\{(h_1 - Z)\sin(\theta_A) + X\cos(\theta_A)\}$	$r_3 = (h_1 - Z)\sin(\theta_A) + X\cos(\theta_A)$
Q_5	$Q_5\{(h_2 - Z)\sin(\theta_C) + (L_5 - X)\cos(\theta_C)\}$	$r_5 = (h_2 - Z)\sin(\theta_C) + (L_5 - X)\cos(\theta_C)$

For sideframe equilibrium, all moments acting about G will sum to zero ($\Sigma M = 0$):

$$-Q_2 X + Q_3\{(h_1 - Z)\sin(\theta_A) + X\cos(\theta_A)\} + Q_4(H - Z) + \cdots$$
$$\cdots Q_5\{(h_2 - Z)\sin(\theta_C) + (L_5 - X)\cos(\theta_C)\} - Q_6(L_5 - X) + \cdots$$
$$\cdots Q_7 Z - Q_{X1}(h_1 - Z) - Q_{X2}(h_2 - Z) = 0$$

Careful inspection of this equation and of Figure 5.14 reveals that the coefficients of the Qs must be their moment arms r_2, r_3, etc., about point G as given in equation (5.27). Moment arms r_2 to r_7 are listed in Tables 5.3(a) and (b).

5.3.5 Summary – baseline closed sedan

Front and rear inner fenders:

$$R_{FT} = T/S_F \qquad R_{RT} = T/S_R = R_{FT}S_F/S_R$$
$$P_{FT} = TL_1/S_F h_1 \qquad P_{RT} = TL_2/S_R h_2$$

Parcel shelves:

$$Q_{X1} = P_{FT}S_F/B = TL_1/h_1 B \qquad (5.19)$$
$$Q_{X2} = P_{RT}S_R/B = TL_2/h_2 B \qquad (5.20)$$

Passenger compartment (with some re-arrangements):

$$Q_1 h_1 + Q_2 B = T \qquad\qquad (5.21) \quad \text{front bulkhead}$$

$$-Q_1 h_3 + Q_3 B = 0 \qquad\qquad (5.22) \quad \text{windshield}$$

$$-Q_1 L_7 + Q_4 B = 0 \qquad\qquad (5.23) \quad \text{roof}$$

$$-Q_1 h_4 + Q_5 B = 0 \qquad\qquad (5.24) \quad \text{backlight}$$

$$Q_1 h_2 + Q_6 B = T \qquad\qquad (5.25) \quad \text{rear bulkhead}$$

$$Q_1 L_5 - Q_7 B = P_{FT} S_F + P_{RT} S_R \qquad (5.26) \quad \text{floor}$$

$$-r_2 Q_2 + r_3 Q_3 + r_4 Q_4 + r_5 Q_5 - r_6 Q_6 + r_7 Q_7$$
$$= Q_{X1}(h_1 - Z) + Q_{X2}(h_2 - Z) \qquad (5.27) \quad \text{sideframe}$$

The input moment is T and P_{FT}, P_{RT}, Q_{x1} and Q_{x2} are known in terms of T. The remaining seven unknown edge forces (Q_1 to Q_7) can be solved using the seven simultaneous equations (5.21 to 5.27) if T is known. This could be done by hand (the equations are linear and relatively 'sparse'). Alternatively, if the equations are rearranged slightly and put in matrix form, they become:

$$
\begin{bmatrix}
L_5 & 0 & 0 & 0 & 0 & 0 & -B \\
h_1 & B & 0 & 0 & 0 & 0 & 0 \\
-h_3 & 0 & B & 0 & 0 & 0 & 0 \\
-L_7 & 0 & 0 & B & 0 & 0 & 0 \\
-h_4 & 0 & 0 & 0 & B & 0 & 0 \\
h_2 & 0 & 0 & 0 & 0 & B & 0 \\
0 & -r_2 & r_3 & r_4 & r_5 & -r_6 & r_7
\end{bmatrix}
\begin{bmatrix}
Q_1 \\ Q_2 \\ Q_3 \\ Q_4 \\ Q_5 \\ Q_6 \\ Q_7
\end{bmatrix}
=
\begin{bmatrix}
P_{FT} S_F + P_{RT} S_R \\
T \\
0 \\
0 \\
0 \\
T \\
Q_{X1}(h_1 - Z) \\ +Q_{X2}(h_2 - Z)
\end{bmatrix}
$$

(5.26) floor
(5.21) front bulkhead
(5.22) windshield
(5.23) roof
(5.24) backlight
(5.25) rear bulkhead
(5.27) sideframe

Equilibrium matrix   Edge forces   Input forces

They can then be solved using a standard computer method (e.g. Gaussian reduction) available widely. Most 'spreadsheet' programs can do this. To ensure reliability of the solution, the matrix should be 'positive definite'. This requires all terms on the leading diagonal of the coefficient matrix to be positive and non-zero, and this is the reason for the rearrangement of the order in which the equations are listed.

Solution of the equations gives values for the edge forces Q_1 to Q_7.

The stresses in the shear panels and ring frames can then be estimated. This is discussed in Chapters 7 and 11.

Solution check

It is easy to make errors in the setting up of the equilibrium equations. The solution for edge forces Q should be checked to ensure force balance in both horizontal and vertical directions on the sideframe. Referring to Figures 5.14 and 5.15:

$$\sum F_{\text{HORIZONTAL}} = 0 \ldots Q_{3\ \text{HORIZONTAL}} + Q_4 + Q_{5\ \text{HORIZONTAL}} - Q_7 - Q_{X1} - Q_{X2} = 0$$

$$\sum F_{\text{VERTICAL}} = 0 \ldots - Q_2 + Q_{3\ \text{VERTICAL}} - Q_{5\ \text{VERTICAL}} + Q_6 = 0$$

5.3.6   Some notes on the standard sedan in torsion

(a) A knowledge of the shear stress τ in the panels is required. The panel thickness t must be chosen so that shear stress is within permitted values. The average shear stress τ in the continuous shear panels can be calculated from the edge force Q by:

$$\tau_{\text{AV}} = \frac{Q}{\text{shear area}}$$

The force intensity on a panel edge is often expressed as shear force per unit length or 'shear flow' q. Shear flow is related to shear stress thus $q = t\tau$. As with shear stress systems, every shear flow has an equal 'complementary' shear flow q' at $90°$ to it.

For example in the roof (see Figure 5.12):

Shear flow on front of roof $q_1 = Q_1/B$. Shear flow on side of roof $q_4 = Q_4/L_7$ but the latter would be the complementary shear flow to q_1. From equation (5.23) above:

$$Q_1 L_7 = Q_4 B$$

hence, substituting: $$(q_1 B)L_7 = (q_4 L_7)B$$

thus $$q_1 = q_4 \quad \text{(complementary shear flows)}$$

For a car of constant width B such as this one, the shear flows q_1 on panel edges across the car are all the same at $q_1 = Q_1/B$. For the panels subject to edge forces only (windshield, roof, backlight in this case) the complementary shear flows $q_3\ q_4\ q_5$ will be equal to this, and hence equal to each other. The shear flow around this part of the sideframe is thus constant. This reflects the Bredt–Batho theorem for shear flows in closed sections subject to torsion, because the passenger compartment can be thought of as a closed tube running across the car between the sideframes. For some of the other panels, such as bulkheads and floor, the edge forces calculated above are affected by additional external moments on the panel (e.g. $R_{\text{TF}}S_F$ on the front bulkhead). This masks the complementary shear flow effect in these cases. For example, the couple $T = R_{\text{FT}}S_F$ is introduced into the front bulkhead by forces R_{FT} (see Figure 5.12) so that the shear force (and hence the local shear flow along the top and bottom edges) varies across the panel. Thus, force Q_1 is the *net* force on the top and bottom of this bulkhead. (This is discussed further in section 7.2.5.)

The equality of complementary shear flows can also be used as a check on the solutions for the forces Q on the structural surfaces that have edge forces only (e.g. roof, windshield). Again, this check cannot be applied directly to structural surfaces (e.g. floor) which are subject to extra forces in addition to the edge forces.

(b) The 'shear panel' load path in the compartment depends on all simple structural surfaces in it carrying shear effectively.

The least effective SSSs in this respect are the ring frames, including the sideframes and the windshield frame. This type of structural surface achieves its in-plane shear

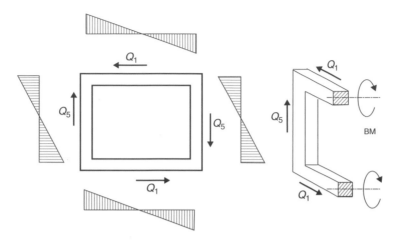

Figure 5.16 Ring beam.

stiffness by acting as a ring beam, with local bending of its edge beams about an axis normal to the plane of the frame. As discussed earlier, this leads to high local bending moments at the corners of the frame (see Figure 5.16). The edge members are said to be in 'contraflexure' since the bending moment (and hence the curvature) changes sign part-way along the member.

The least effective of these frames is the slender upper A-pillar. Thickening the lower A-pillar will cause more of Q_{X1} etc., to be conveyed downward to the underbody (so-called 'semi-open' structure).

If the windshield glass is adhesive bonded to the frame, then the glass will act as a substantial load path (as a shear panel). This will relieve the corner bending moments on the frame (somewhat) and will make a significant increase in the shear stiffness of the window frame. Cars with bonded screens have shown large torsion stiffness increases (up to 60%). Care must be taken, however, to ensure that the load in the glass does not cause it to break in extreme load cases (e.g. vehicle corner bump case, or wheel jacking for puncture). The vehicle stiffness and member stresses (e.g. corner bending moment in window frame) should not degrade to unacceptable values in the event of a smashed windshield.

Windshields attached with (elastomer) gaskets give a much less significant contribution to the structure, due to the flexibility of the gasket.

(c) The sideframe consists of two (or sometimes three) rings, bordered by the A-, B- and C- (etc.) pillars. The share of the shear load Q_4 carried by each of these beams will be in proportion to their relative stiffness and hence will depend on their second moments of area, and lengths (see Figure 5.17).

Calculating the share of the load in each pillar and hence the bending moments and the stresses in the ring frames accurately is complicated, particularly in the multiple ring case (sideframe), since they are statically indeterminate (three redundancies per 2D ring). Approximate methods of estimating these are discussed in Chapter 7.

The sideframe is subject to overall shear, and since it is a (multi-bay) ring beam, its edge members experience 'contraflexure bending', giving high joint stresses as shown in Figure 5.17.

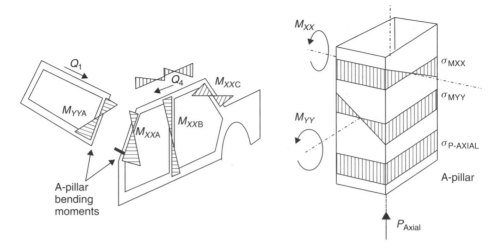

Figure 5.17 Stresses in shared pillar.

(d) As noted above, the ring frames have high bending moments at their corners. Some pillars are shared by two adjacent ring frames (e.g. the upper A-pillar is shared by the windshield surround and sideframe. See Figure 5.17). In this case the lateral load *in each direction* P_X and P_Y must be accounted for. They will lead to bending moments about two directions in the pillar. If the pillar is sloped in one of these frames (e.g. the A-pillar in the sideframe) then there will also be an axial component of force in the member. These moments and axial forces will cause a combined stress system in the pillar.

$$\sigma_{\max.} = \frac{M_x y}{I_x} + \frac{M_y x}{I_y} + \frac{P}{A}$$

This is discussed further in Chapter 7.

(e) A good load path from the inner fender upper booms to the sideframes, via the 'parcel shelf' is essential (unless the booms are offset to connect directly to the sideframe A-pillar). Thus the parcel shelf must be stiff in bending. It is a highly stressed component because of the high bending moments caused by forces P_{FT}.

(f) The connection of the parcel shelf to the A-pillar is another high stress area. This results from the high local bending moment in the A-pillar due to force Q_{x1}. Similarly, in the standard sedan, the middle of the C-pillar will be highly stressed due to bending moments caused by the 'hat shelf' end reactions Q_{x2}.

5.3.7   Structural problems in the torsion case

The 'faux sedan'

A 'closed box' of simple structural surfaces is required in the passenger compartment to maintain an effective load path of 'shear panels' to carry torsion.

If any one of the structural surfaces in the compartment is missing then the structure becomes an 'open box' and the shear panel load path for torsion breaks down.

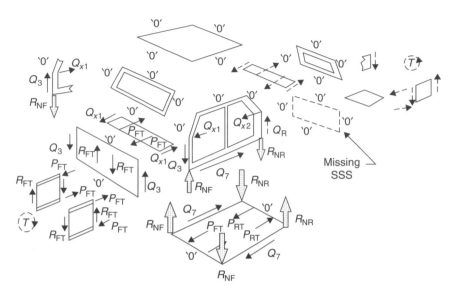

Figure 5.18 Faux sedan in torsion.

For example, sometimes the rear seat bulkhead is omitted to allow a 'split rear seat' feature (see Figure 5.18). In this case there is no ability to react edge forces Q_1 and Q_6 around the rear seat back, so that the shear forces Q_1 at the bottom of the rear window and on the rear of the floor must be zero. Hence (for force equilibrium) the edge forces Q_1 at the other ends of these surfaces become zero. This will 'propagate' all the way round the compartment so that all forces Q_1 (roof, windshield, front bulkhead, floor) must also be zero. The complementary forces on all of these structural surfaces will also become zero.

The only compartment edge forces remaining will be those which can be balanced by 'external' torques. Thus in the front bulkhead the torque T is balanced by the couple Q_2B from forces Q_2 at the A-pillar. The couple Q_7B from edge forces Q_7 on the floor are balanced by the couples $P_{FT}S_F$ and $P_{RT}S_R$.

It is assumed here that a load path is available to transmit the reaction torque T (from the rear suspension) to the sideframe as edge forces Q_R on the C-pillars via the *outer* rear fender and/or via floor members. (The rear seat bulkhead is no longer available for this.)

The moments on the sideframe due to forces $Q_{x1} Q_{x2} Q_2 Q_R$ and Q_7 (see Figure 5.18) are all in the same direction. These moments can only be balanced by moments from forces R_{NF} and R_{NR}. These cause reactions R_{NF}, R_{NR} *normal to the plane of the floor*. This causes the floor to twist out-of-plane. The SSS assumptions are not satisfied in this case. Similar twisting of other surfaces (e.g. front bulkhead/parcel shelf) will occur.

In practice, such a 'faux sedan' structure is much more flexible (and less weight efficient) in torsion than a 'closed box' sedan. The sideframes tend to act as 'levers' to twist the cowl/dash (engine bulkhead/parcel shelf) assembly.

High local stresses and large strains are experienced by the remaining loaded members (in the floor etc.) leading to: (a) poor fatigue life and (b) damage to the paint system with resulting early corrosion.

The same 'faux sedan' problem will be encountered if any one (or more) of the compartment structural surfaces are missing or of reduced structural integrity (e.g. roof, front bulkhead or floor missing, or poorly connected to adjacent components). Similarly loss of integrity of sideframe 'ring beam' members: e.g. poor quality joints, 'panoramic' A-pillars, pillarless sedans (to some extent), corroded rockers, etc. – especially if the weakening is at the high bending moment corners of the ring frames.

Remedies for the faux sedan

Replacement of missing shear panel with ring frame

The ideal remedy for the faux sedan is to modify it so as to restore the 'closed box' type structure with its weight efficient shear panel load paths.

This is possible, in some cases, by using a ring frame in place of the missing panel (Figure 5.19). This restores shear integrity to the surface in question, whilst maintaining a substantial opening. For such a ring frame, care must be taken to ensure an effective path for local bending moments all round the frame *especially* at the corners. For example, in the rear seat bulkhead case, parcel shelf, side-wall beams and floor cross-member (which could be under the floor) must all have high stiffness for bending about axes normal to the plane of the frame, *and* they *must all* be well connected for bending at the corners (e.g. gusseted joints). C-pillars and parcel shelves are often not stiff in the required direction, and so require special design attention.

If a larger opening is required (e.g. hatchback, station wagon, opening rear screen with split seat) then a ring frame, running the full height of the sideframe (through the C-pillars, across the roof and across the floor), is a possibility. This works better if the ring frame is as planar as possible, and if the corner joints are good in bending. Even so, the result will not be as weight efficient as a true shear panel.

An alternative to the ring beam is a triangulated bay in the opening (see Figure 5.13).

Provision of 'closed torque box' in part of structure

There are several areas in the integral car body that are 'box like'. Some of these are:

(a) The cowl/footwell assembly (the region enclosed by the parcel shelf, the engine bulkhead, the lower A-pillars and the floor).

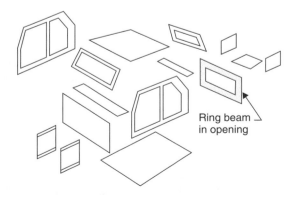

Ring beam in opening

Figure 5.19 Ring frame to remedy faux sedan problem.

(b) The engine compartment and/or the rear luggage compartment.

(c) The region under the rear seat, where there is often a step in the floor.

If any of these can be converted into a 'closed box' structure by the addition of shear panels or ring frames, *and* can be well connected to the sideframes, then this will provide some torsion stiffness. The sideframes, acting as stiff levers, will make this torsion stiffness available to the whole body. The added members making up the box may already be present, providing other functions (e.g. instrument panel, floor cross-member, etc.). Examples are shown in Figure 5.20.

This approach is likely to be less weight efficient, and more prone to high stress/strain problems (fatigue, corrosion) than the true 'closed' integral structure described in section 5.3.4. (See section 6.4.4 for further discussion.)

Provision of true grillage structure in floor

A grillage is a flat structure which is stiff (for out-of-plane loads) in both bending and twist. Individual members in a true grillage experience local bending moments and torques (both about axes within the plane of the grillage) and shear forces normal to the plane. They must therefore be stiff for these loads (see Figure 5.21).

Hence, they are likely to be closed section (box) members of considerable depth. There are a few versions of grillage structures with torsion stiffness for which individual members require only bending stiffness (e.g. cruciform grillages, see Chapter 7).

A degree of torsion stiffness could be restored to the faux sedan by building a torsionally stiff grillage into the floor. This is not very weight efficient. The result is called a 'semi-open' structure since structurally it behaves like an open ('convertible') car structure of the 'punt' type, the upper structure contributing little to the torsional performance. See the discussion of convertible and punt structures in Chapters 3, 6 and 7.

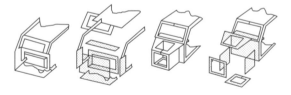

Figure 5.20 Torsion stiffening by 'boxing in' localized regions of body.

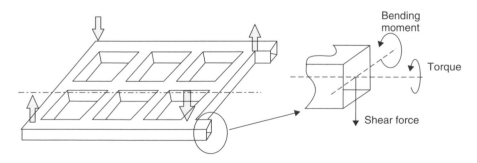

Figure 5.21 Grillage structure.

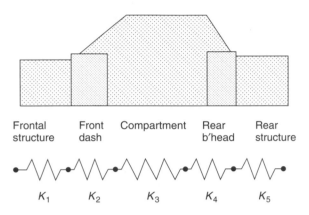

Frontal structure	Front dash	Compartment	Rear b'head	Rear structure
K_1	K_2	K_3	K_4	K_5

Figure 5.22 Structures in series.

Effect of one poor subassembly on overall torsion stiffness

For the torsion case, the vehicle behaves as a set of five subassemblies in series in torsion as shown in Figure 5.22 (1. Frontal structure, 2. Dash, 3. Passenger compartment, 4. Rear bulkhead/hat shelf, 5. Rear structure).

For structures in series, the overall torsion stiffness K is given by:

$$1/K = 1/K_1 + 1/K_2 + 1/K_3 + 1/K_4 + 1/K_5$$

where K_1 to K_5 are the stiffnesses of the assemblies listed above.

In such a series, the overall flexibility $(1/K)$ is dominated by any member that has very low stiffness. For example, consider a vehicle where K_1 to K_5 all have the value 50 000 Nm/deg. Thus in this case:

$$1/K = 5/50\,000, \text{ hence overall torsion stiffness: } K = 10\,000 \text{ Nm/deg.}$$

Now suppose that the dash has reduced torsional stiffness, $K_2 = 5000$ Nm/deg. This might be because of a faulty load path (for example, insufficiently stiff parcel shelf). The causes of poor dash performance in torsion are very similar to those in the vehicle bending case (see discussion in section 5.2.5). The overall torsion stiffness of the vehicle is now given by:

$$1/K = 1/K_2 + (1/K_1 + 1/K_3 + 1/K_4 + 1/K_5) = 1/5000 + 4/50\,000$$

Thus: $K = 3571$ Nm/deg.

The overall torsion stiffness has been reduced to about 36% of the original value. It is thus of paramount importance that all of the subassemblies have correct load-path design and that the connections between them are structurally sound.

5.4   Lateral loading case

When a vehicle travels on a curved path lateral forces are generated due to centrifugal acceleration. Inertia forces act at the centres of mass of the components which tend

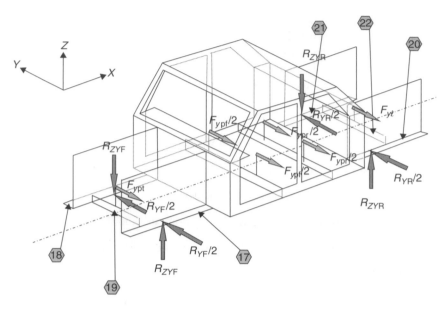

Figure 5.23 Baseline model – lateral loading with additional SSSs.

to throw them away from the centre of turn. These are balanced by lateral forces generated at the tyre to ground contact points which are transferred to the structure of the vehicle through the suspension. This condition is illustrated in Figure 5.23 where the vehicle is considered moving forward and turning to the right.

The inertia forces on the power-train, the front and rear passengers and the luggage are shown acting towards the left of the vehicle. The balancing side forces are shown as R_{YF} and R_{YR} acting at the front and rear axles, respectively. As the lateral forces at the centres of mass act above the floor line these produce a rolling moment that is balanced by vertical loads R_{ZYF} and R_{ZYR} at the front and rear suspension mountings. From Figure 5.23, it should be noted that these forces act downward on the right-hand side and upward on the left-hand side. These forces are sometimes known as the weight transfer due to cornering. It is important to realize that these are forces which act in *addition* to the vertical forces shown in Figure 5.2. In this section only these forces are considered and they should then be added to the force system analysed in section 5.2.

5.4.1 Roll moment and distribution at front and rear suspensions

Taking moments about the vehicle centreline (see Figure 5.24) at the plane of the floor to obtain:

$$\text{Roll moment} \quad M_R = F_{ypt}h_{pt} + F_{ypf}h_{pf} + F_{ypr}h_{pr} + F_{y\ell}h_\ell$$

This is the moment due to the forces acting through the centres of mass. This is balanced by the vertical reactions at front and rear suspension mounting points:

$$M_R = R_{ZYF}t_f + R_{ZYR}t_r$$

There are now unknowns, R_{ZYF} and R_{ZYR}, and only one equation. A ratio must be assumed between these two unknowns. The roll stiffness of the front suspension of most cars is greater than the roll stiffness of the rear suspension and the body roll angle at front and rear is assumed equal (the body is very stiff compared to the suspension), therefore it will be assumed that the front suspension provides nM of the roll moment, where n is in the range 0.5 to 0.7.

Therefore at the front suspension mounting:

$$R_{ZYF} = nM_R/t_f$$

and at the rear suspension mounting:

$$R_{ZYR} = (1 - n)M_R/t_r$$

Returning to Figure 5.24 and the plan view, take moments about the front suspension:

$$R_{YR} = \{F_{ypf}(l_1 + l_3) + F_{ypr}(l_1 + l_4) + F_{y\ell}(L + l_\ell) - F_{ypt}l_{pt}\}/L$$

Resolving lateral forces:

$$R_{YF} = \{F_{ypt} + F_{ypf} + F_{ypr} + F_{y\ell}\} - R_{YR}$$

The magnitude of the lateral forces is discussed in section 2.4.4 therefore the loads on the structure can be evaluated.

5.4.2 Additional simple structural surfaces for lateral load case

As the lateral forces act through the centres of mass of the power-train and the luggage and are in front of and behind the centre floor additional SSSs (17) to (22) are required to those shown in Figure 5.1 and are shown in Figure 5.23. SSS (19) will transfer the

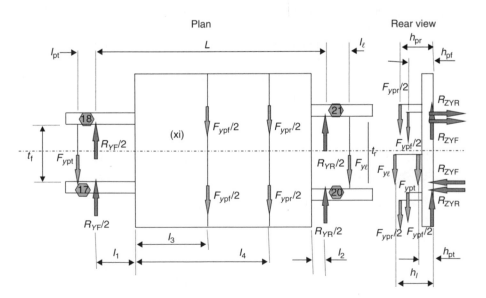

Figure 5.24 Baseline model – lateral loading.

power-train force as vertical forces in the planes of the front inner wing panels and it will be assumed the lateral force is shared equally between SSS (17) and SSS (18) (see Figure 5.24). Similar conditions are assumed to act at the luggage floor beam. As the beams (19) and (22) cannot take moments about the vehicle z-axis, the beams (17), (18), (20) and (21) act as simple cantilevers protruding forward and rearward from the central floor.

Consider the cross-beams all shown in Figure 5.25

Engine beam (19)

By taking moments about one end $\qquad P'_{14} = F_{ypt}h_{pt}/t_f \qquad$ (5.28)

Resolve forces laterally $\qquad\qquad P'_{15} = F_{ypt}/2 \qquad\qquad$ (5.29)

Transverse floor beam (front) (1)

By taking moments about one end $\qquad P'_1 = F_{ypf}h_{pf}/w \qquad$ (5.30)

Resolving forces laterally $\qquad\qquad P'_{16} = F_{ypf} \qquad\qquad$ (5.31)

Transverse floor beam (rear) (2)

By taking moments about one end $\qquad P'_2 = F_{ypr}h_{pr}/w \qquad$ (5.32)

Resolving forces laterally $\qquad\qquad P'_{17} = F_{ypr} \qquad\qquad$ (5.33)

Luggage beam (22)

By taking moments about one end $\quad P'_{18} = F_{y\ell}h_{\ell}/t_r \qquad$ (5.34)

Resolving forces laterally $\qquad\qquad P'_{19} = F_{y\ell}/2 \qquad\qquad$ (5.35)

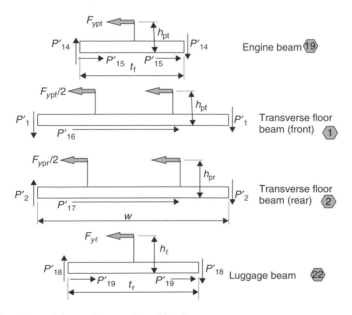

Figure 5.25 Baseline model cross-beams – lateral loading.

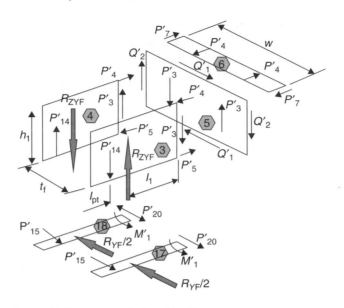

Figure 5.26 Baseline model (front structure) – lateral loading.

Now moving to the front structure around the engine compartment (Figure 5.26), consider each SSS.

Lower rails of front inner panels (17 and 18)

By taking moments about the rear end

$$M_1 = R_{YF}l_1/2 - P'_{15}(l_1 + l_{pt}) \tag{5.36}$$

Resolve forces laterally

$$P'_{20} = R_{YF}/2 - P'_{15} \tag{5.37}$$

Left-hand front inner wing panel (3)

Resolve forces vertically

$$P'_3 = R_{ZYF} - P'_{14} \tag{5.38}$$

By taking moments about lower rear corner

$$P'_4 = \{R_{ZYF}l_1 - P'_{14}(l_1 + l_{pt})\}/h_1 \tag{5.39}$$

Resolve forces horizontally

$$P'_5 = P'_4 \tag{5.40}$$

The right-hand front inner wing panel has similar load values but all are in the opposite sense giving similar equations.

Front parcel shelf (6)

This member is attached to the dash panel and the windscreen frame at the front edge but with no attachment at the rear. Therefore, no lateral loads can be applied. The forces from the inner wing panels P'_4 are reacted by the forces P'_7 acting on the sideframe.

By taking moments

$$P'_7 w - P'_4 t_f = 0 \tag{5.41}$$

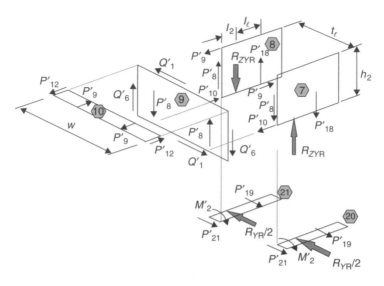

Figure 5.27 Baseline model (rear structure) – lateral loading.

The rear structure is shown in Figure 5.27 and has similar loading conditions to the front structure.

Lower rails of rear quarter panels (20 and 21)

Resolve forces laterally $P'_{21} = R_{YR}/2 - P'_{19}$ (5.42)

By taking moments about the front end

$$M'_2 = R_{YR}l_2/2 - P'_{19}(l_2 + l_\ell)$$ (5.43)

Left-hand rear quarter panel (7)

Resolve forces vertically $P'_8 = R_{ZYR} - P'_{18}$ (5.44)

By taking moments about lower front corner

$$P'_9 = \{R_{ZYR}l_2 - P'_{18}(l_2 + l_1)\}/h_2$$ (5.45)

Resolve forces horizontally $P'_{10} = P'_9$ (5.46)

The right-hand rear quarter panel (8) has opposite forces acting on it and these will produce identical equations.

Rear parcel shelf (10)

Similar to the front parcel tray, no lateral forces act on this member.

By taking moments $P'_{12}w - P'_9 t_r = 0$ (5.47)

By working through the equations (5.28) to (5.47) in the sequence shown, all the unknowns (i.e. $P'_1 \ldots P_5$, $P'_7 \ldots P'_{10}$, P'_{12}, $P'_{14} \ldots P'_{21}$, M'_1 and M'_2) can be evaluated.

To evaluate the forces in the passenger compartment shown in Figure 5.28 consider the seven SSSs forming the 'torsion box'. The right-hand sideframe is not shown as

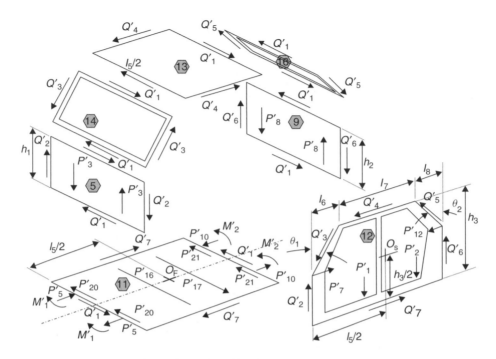

Figure 5.28 Baseline model (torsion box) – lateral loading.

its loading condition is exactly opposite to the left-hand sideframe. It should be noted that the moments applied into the 'torsion box' via the dash panel (5) and the panel behind the rear seats (9) are not in the opposite sense as in the torsion case described in section 5.3. These two moments in this case are unequal and in the same sense, because they both contribute to balancing the moments due to the moments applied through the floor cross-beams. The effect of these unequal moments is to apply shear to all the SSSs forming the 'torsion box'.

Dash panel (5)
The loads P'_3 from the inner wing panels produce a moment that is balanced by edge loads at the sides, top and bottom. These loads Q'_1 and Q'_2 are assumed to act as shown.

$$\text{Moment equation} \qquad Q'_1 h_1 + Q'_2 w - P'_3 t_f = 0 \qquad (5.48)$$

Windscreen frame (14)
This member must react the edge load Q'_1 from the dash panel and is held in equilibrium by the complementary shear forces Q'_3.

$$\text{Moment equation} \qquad Q'_1 (l_6^2 + (h_3 - h_1)^2)^{0.5} - Q'_3 w = 0 \qquad (5.49)$$

Roof (8)
This again is in complementary shear.

$$\text{Moment equation} \qquad Q'_1 l_7 - Q'_4 w = 0 \qquad (5.50)$$

Backlight frame (16)

Once again this is in complementary shear.

Moment equation $Q'_1(l_8^2 + (h_3 - h_2)^2)^{0.5} - Q'_5 w = 0$ (5.51)

Panel behind the rear seats (9)

Loads P'_8 applied from the rear quarter panels apply a moment to this panel. Edge load Q'_1 is applied to the top edge from the backlight frame, therefore an equal and opposite force Q'_1 acts at the lower edge. All these forces apply moments in the same sense, therefore to maintain equilibrium Q'_6 at each side must act in the directions shown.

Moment equation $Q'_6 w - Q'_1 h_2 - P'_8 t_r = 0$ (5.52)

Floor panel (11)

There are a large number of loads acting on this SSS. Shear loads, tension/compression loads and moments act on the front and rear edges plus shear loads from the front and rear transverse floor beams. In order to achieve equilibrium, loads Q'_7 are required at the sides.

Moments about the centre of the floor (O_F)

$$(2P'_{20} - Q'_1)l_5/2 + P'_5 t_f + 2M'_1 - (2P'_{21} + Q'_1)l_5/2 - 2M'_2$$
$$- P'_{10} t_r + P_{16}(l_3 - l_5/2) + P_{17}(l_4 - l_5/2) + Q'_7 w = 0 \qquad (5.53)$$

Left-hand sideframe (12)

The forces acting on this SSS are shown (Figure 5.28). These are the equal and opposite reactions to the loads on the previous six SSSs plus loads from the front and rear parcel shelves and from the transverse floor beams.

Moments about the centre of the sideframe (O_S)

$$Q'_2 l_5/2 + P'_1(l_3 - l_5/2) + P'_2(l_4 - l_5/2) - Q_6 l_5/2) + P'_7(h_1 - h_3/2)$$
$$- P'_{12}(h_2 - h_3/2) - Q_4 h_3/2 - Q_7 h_3/2 - Q'_3(h_3/2)\sin\theta_1 - Q'_3(l_5/2 - l_6)\cos\theta_1$$
$$- Q'_5(h_3/2)\sin\theta_2 - Q'_5(l_7 + l_6 - l_5/2)\cos\theta_2 = 0 \qquad (5.54)$$

Equations (5.48) to (5.54) are seven simultaneous equations with seven unknowns $Q'_1 \ldots Q'_7$, so their solutions can be obtained by standard mathematical methods. Equation (5.54) can be simplified by taking moments about the lower corner of the windscreen pillar. This will eliminate terms in Q'_2, Q'_3 and P'_7 but the numerical solutions will be unchanged. These equations provide the values of all the forces on the SSSs of the 'torsion box'. When evaluating these forces accidental errors may occur, therefore it is *imperative* that the sideframe equilibrium is verified. This is done by resolving forces vertically and horizontally using the following equations:

Resolve forces vertically

$$Q'_2 - Q'_3 \cos\theta_1 + Q'_5 \cos\theta_2 + Q'_6 - P'_1 - P'_2 = 0 \qquad (5.55)$$

Resolve forces horizontally

$$P'_7 - P'_{12} + Q'_7 - Q'_3 \sin\theta_1 - Q'_4 - Q'_5 \sin\theta_2 = 0 \qquad (5.56)$$

Provided these equations are satisfied, the correct force conditions through the structure have been determined.

It should be observed that lateral loads cause additional bending on the front inner wing panels, rear quarter panels, parcel shelves and shear on the dash panel, windscreen frame, roof, backlight frame, and seat panel. The main floor and the sideframes have both additional shear and bending. These additional loads must be added by superposition to those obtained in section 5.2.

5.5   Braking (longitudinal) loads

A passenger car subject to braking conditions has additional loads shown in Figure 5.29 over the normal bending loads. The proportion of the total braking force applied at the front wheels is usually in the order of 50 to 80% (Newcomb and Spurr 1966). This is due to the design of the braking system and because additional vertical load R_{ZXF} is applied to the front wheels. Modern braking systems may have variable brake proportioning to suit the decelerating condition. For this analysis the assumed proportion of braking on the front axle is:

$$n = \frac{R_{XF}}{R_{XF} + R_{XR}} \tag{5.57}$$

Forces R_{XF} and R_{XR} are applied at the front and rear tyre to ground contacts which are h_f below the base of the structure. The load transfer onto the front axle and off the rear axle is obtained by taking moments about the rear tyre/ground contact point (see Figure 5.30):

$$R_{ZXF} = R_{ZXR} = \{F_{xpt}(h_{pt} + h_f) + F_{xpf}(h_{pf} + h_f) + F_{xpr}(h_{pr} + h_f) + F_{x\ell}(h_\ell + h_f)\}/L \tag{5.28}$$

The loading on the front structure (Figure 5.31) is now known so applying the equations of statics to each of the SSSs.

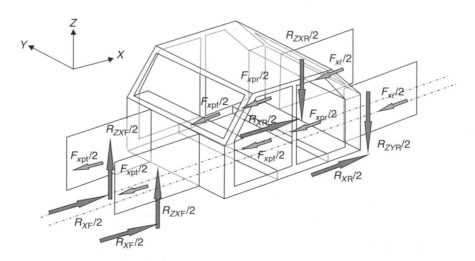

Figure 5.29 Baseline model – braking loads.

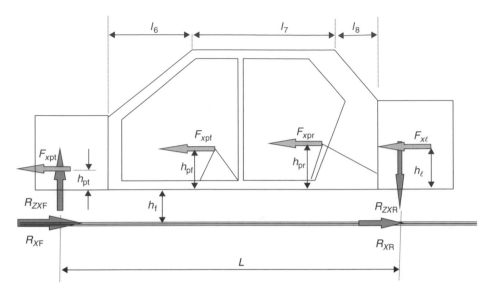

Figure 5.30 Baseline model (side view) – braking loads.

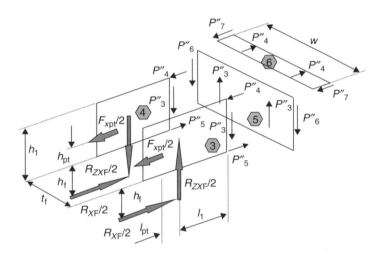

Figure 5.31 Baseline model (front structure) – braking loads.

Inner wing panels (3) and (4)

The inertia forces from the power-train $F_{xpt}/2$ are assumed to act into each of the inner wing panels.

Resolve forces vertically	$R_{ZXF}/2 - P_3'' = 0$	(5.29)
Resolve forces horizontally	$R_{XF}/2 - F_{xpt}/2 - P_4'' + P_5'' = 0$	(5.30)

Moments about the lower rear corner

$$R_{ZXF}l_1/2 - F_{xpt}h_{pt}/2 - R_{XF}h_f/2 - P_4''h_1 = 0 \qquad (5.31)$$

Hence

$$P_3'', P_4'' \text{ and } P_5'' \quad \text{are obtained.}$$

Dash panel (5)
Resolving forces vertically and by symmetry

$$2P_6'' - 2P_3'' = 0 \tag{5.32}$$

Front parcel shelf (6)
Resolving forces horizontally $\quad 2P_7'' - 2P_4'' = 0$ $\tag{5.33}$

The rear structure shown in Figure 5.32 is loaded such that the suspension and luggage loads all act on the rear quarter panel.

Rear quarter panels (7) and (8)
Resolving forces vertically $\quad\quad\quad P_8'' - R_{ZXR}/2 = 0$ $\tag{5.34}$

Resolving forces horizontally $\quad\quad P_{10}'' + R_{XR}/2 - P_9'' - F_{x\ell}/2 = 0$ $\tag{5.35}$

Moments about front lower corner

$$R_{ZXR}l_2/2 - R_{XR}hf/2 - F_{x\ell}h_\ell/2 - P_9''h_2 = 0 \tag{5.36}$$

Panel behind rear seat (9)
This member provides reaction for the shear force P_8'' and an additional force P_{26}''. P_{26}'' is caused by the inertia load of the rear passengers

$$P_{26}''(l_5 - l_4) - F_{xpr}h_{pr}/2 = 0 \tag{5.37}$$

Resolving forces vertically and by symmetry

$$P_{11}'' + P_{26}'' - P_8'' = 0 \tag{5.38}$$

Rear parcel shelf (10)

Resolving forces horizontally $\quad 2P_{12}'' - 2P_9'' = 0$ $\tag{5.39}$

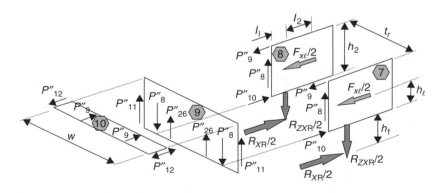

Figure 5.32 Baseline model (rear structure) – braking loads.

Transverse floor beam (rear) (2)

As the inertia load from the rear seat passengers acts at height h_{pr} the vertical load on this member is:

$$P''_{25} = F_{xpr}h_{pr}/2(l_5 - l_4) \tag{5.40}$$

Transverse floor beam (front) (1)

Again the inertia load from the front passengers acts at a height h_{pf} resulting in a moment about the floor plane. In this situation an extra floor beam is required and so two front floor beams as shown with dashed lines in Figure 5.33 are used to replace the original one:

$$P''_{22} = P''_{23} = F_{xpf}h_{pf}/2l_{10} \tag{5.41}$$

Floor panel (11)

The inertia forces from the front and rear passengers are transferred as shear to the floor:

$$P''_{24} = F_{xpf}/2 \tag{5.42}$$

$$P''_{27} = F_{xpr}/2 \tag{5.43}$$

Resolving forces horizontally and by symmetry

$$P''_{13} = P''_5 + P''_{10} + P''_{24} + P''_{27} \tag{5.44}$$

Finally, once again check the equilibrium of the sideframe by the three equations of statics.

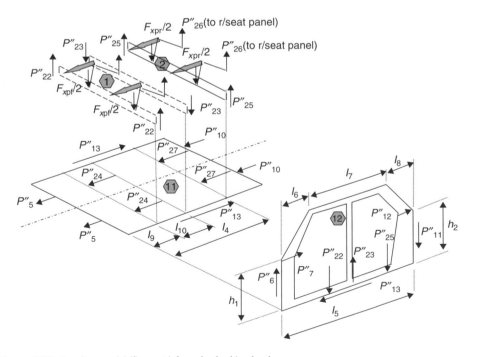

Figure 5.33 Baseline model (floor, sideframe) – braking loads.

Resolving forces horizontally $P_7'' + P_{12}'' - P_{13}'' = 0$ (5.45)

Resolving forces vertically $P_6'' - P_{11}'' - P_{22}'' + P_{23}'' - P_{25}'' = 0$ (5.46)

Moments about lower front corner

$$P_7'' h_1 + P_{22}'' l_9 - P_{23}'' (l_9 + l_{10}) + P_{25}'' l_4 + P_{11}'' l_5 + P_{12}'' h_2 = 0$$ (5.47)

5.6   Summary and discussion

The main load conditions that act on a typical passenger car structure have been considered. The important cases that are the most straightforward to model are the bending and torsion cases. Other cases such as the cornering and the braking have also been modelled. Further cases such as the acceleration or tractive force case and the one wheel bump/pot-hole can also be modelled although not detailed here. The acceleration case is similar to the braking case except that the tractive force is only applied to either the front or the back wheels, not both, unless the vehicle is a four wheel drive vehicle. For the four wheel drive vehicle the proportion of tractive force on front and rear wheels, like in the braking case, will not necessarily be equal.

Although only a 'standard sedan' has been considered the analysis contained in this chapter has revealed that when considering the various load cases the SSSs that are required are quite numerous. In all of the cases considered care must be taken to ensure that there are suitable SSSs to carry the applied loads from one component to another. It should be noted that in the torsion case it is easy to find that insufficient SSSs have been specified as in the 'faux' sedan. Likewise the need for additional SSSs to carry lateral loads should be noted.

In modelling a structure when it is found that it is not reasonable to represent a component with an SSS and that the loads are not satisfactorily transferred through the structure then a weakness in the structure is revealed. This is one of the main advantages of the SSS method – it is useful in revealing if the structure has adequate load paths.

The load conditions on the main structural components have also been determined by the methods described in this chapter. The bending moments, shear forces, distributed edge loads and joint loads can be evaluated using the equations developed. In cases such as the A-pillar (windscreen side), the loading is sufficiently detailed that stress calculations can follow directly from this analysis.

The detailed model of any particular vehicle will require the judgement of the structural engineer. In Chapter 8 the modelling of a particular vehicle is shown that has rather different SSSs and this will indicate the care that is necessary in modelling a structure. Reversing that procedure the SSS model can be used in developing the design of a structure. Having obtained suitable load paths through SSSs the designer can then detail subassemblies and components that have the required structural properties.

6

Alternative construction for body subassemblies and model variants

6.1 Introduction

This chapter discusses variations from the basic 'standard sedan' structure at two levels.

1. *Body subassemblies*
   The major subassemblies, particularly at the ends of the vehicle body between the suspension mountings and the passenger compartment, can take many different forms. The type to be chosen for a given application will depend on factors specific to the particular vehicle package. Some of these alternative subassemblies are discussed in section 6.2.
2. *Vehicle model variants*
   A typical vehicle 'model' may well be produced in a range of variants. These might include station wagon, hatchback, convertible, pick-up truck and other versions in addition to the basic sedan. Each of these has structural features which are significantly different from those of the sedan. The structures of some of these different vehicle types are discussed in sections 6.3 and 6.4.

Open-topped vehicles pose a particular set of structural challenges. The trend toward the use of common floor platforms for a range (or even several ranges) of vehicles has become widespread for reasons of economy. This might tempt the vehicle designer to adopt 'lower dominant' structures in which the upper part of the body plays a relatively minor role. For this reason, open vehicles receive special attention in this chapter.

Detailed discussion of rationalization of structures across a vehicle range based on a common platform is deferred to Chapter 9.

It is usually more difficult to provide a satisfactory structure for the vehicle torsion load case. For brevity in this chapter, most examples are discussed principally in relation to the torsion case. It is usually a straightforward matter to 'read over' to the load paths in the vehicle bending case which is the other principal load case (see Chapter 2).

6.2 Alternative construction for major body subunits

In the standard sedan, the end structures carrying the suspension loads R_F and R_R to the passenger compartment are approximated as deep, vertical, 'boom-panel' inner fender assemblies. The shear panel reaction is carried by the front or rear compartment bulkhead, and the flange (or 'boom') reactions are taken by the parcel shelves and the floor. This is a reasonable approximation to the frontal structure in some sedans, where this structure is set inboard to allow sufficient front wheel steering lock. It is not commonly seen at the rear, where a large luggage compartment and small wheel boxes are desirable.

In this section, alternative end structures at front and rear are considered. Most of these are variations on the theme of the standard sedan, so that they are discussed in relation to the explanation of torsion and bending load cases on the standard sedan given in sections 5.2 and 5.3.

Since most of the structures covered in this section act in series with the passenger compartment, it is essential that they are correctly configured structurally. Otherwise, local deficiencies in the end assemblies themselves, or in their connection to the rest of the structure, will lead to a serious degradation of the overall structural performance of the vehicle body, for the reasons given in Chapter 5, section 5.3.7.

(a)   Rear structures

6.2.1   Rear suspension supported on floor beams (Figure 6.1)

For trailing arm and similar rear suspensions, support for suspension spring reactions is sometimes provided by longitudinal floor beams in the floor of the luggage compartment. The member loads in the vehicle torsion case for this arrangement are shown in Figure 6.1.

In both vehicle torsion and bending cases, the suspension load is distributed by the floor beams forward to the rear seat bulkhead, and aft to the rear valance and thence, via the rear quarter panel, to the rear of the sideframe. All the loads are out of plane for the luggage compartment floor, but the floor panel itself is not stiff in this direction.

For the vehicle torsion load case, using the notation in Figure 6.1, forces Q_8 and Q_9 may be calculated by considering moment equilibrium on the floor beams:

$$Q_8 = R_R L_2/(L_1 + L_2) \quad \text{and} \quad Q_9 = R_R L_1/(L_1 + L_2)$$

On the rear valance:

$$Q_{10} = Q_9 S_R/B$$

The top of the luggage compartment often consists only of narrow flanges around the lid opening. The role of this and the floor is only to provide upper and lower flanges for the rear valance and quarter panels. The quarter panel acts as a deep boom-panel cantilever assembly (see Chapter 7), incorporating these flanges as its booms, and with the wheel box as part of the shear panel. At the front of the quarter panel, the boom forces ($Q_{X2} = Q_{10}(L_1 + L_2)/h_2$) and the panel shear force Q_{10} are all reacted by the

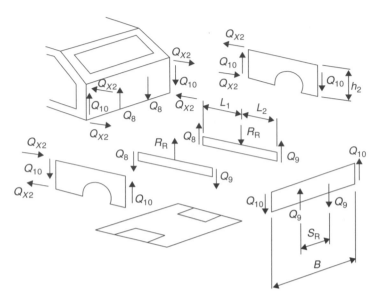

Figure 6.1 Rear floor beam loads (vehicle torsion case).

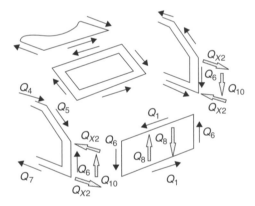

Figure 6.2 Loads applied to the passenger compartment by rear floor beams.

rear of the passenger compartment sideframe. These, and the forces Q_8 on the seat bulkhead, are now all known and can be treated as 'input' forces to the passenger compartment (see Figure 6.2).

Note that the quarter panel shear force Q_{10} is fed directly into the sideframe (instead of into the rear seat bulkhead as in the standard sedan). In addition to this force, the lower C-pillar must also react the rear seat bulkhead edge force Q_6 (see Figures 6.2 and 5.12).

Apart from these small differences in input loads, the edge loads in the compartment are somewhat similar to those in the standard sedan, and they can be solved in much the same way as before (see Chapter 5, section 5.3).

The vehicle bending load case for this structure may be dealt with similarly, except that now all of the reaction forces Q_8 and Q_9 are in the same direction.

6.2.2   Suspension towers at rear

Some vehicles have strut type rear suspension. This requires rear suspension towers, and these are often incorporated into the rear wheel boxes. If these in turn are closely built into the sideframe, then the overall effect is similar to the direct application of the strut reaction into the sidewall, as shown in Figure 6.3. In some designs, a separate structural surface is connected from near the inboard edge of the wheel box to the sideframe to give a more direct load path into the sideframe.

Sometimes, the suspension geometry is such that the line of action of the strut force lies well inboard of the plane of the side of the vehicle (Figure 6.4). The offset between this force and its reaction at the sideframe results in a net couple $R_R B$ on the wheel box/suspension tower. This causes torsion on the sideframe. For the vertical symmetrical vehicle load case this can be mitigated by running a brace across the vehicle between the tops of the suspension towers, as shown in Figure 6.4. Tension in this member and compression in the floor will then balance the moments on the suspension towers. In the vehicle torsion case, the forces on each side of the brace are in the same direction, so that a structure stiff in shear (for example, a shear panel or a ring beam) is needed to react the net force on the top brace, and to carry this reaction to the vehicle floor. With appropriate local design, the brace and the suspension towers can themselves be incorporated into the required ring beam.

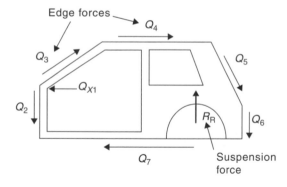

Figure 6.3 Rear suspension tower force acting on body sideframe.

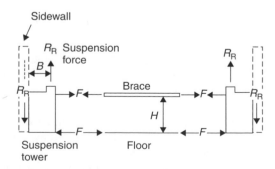

Figure 6.4 Brace to balance moment from offset suspension force.

(b)   Frontal structures

6.2.3   Grillage type frontal structure (see Figure 6.5)

If the front suspension is attached to the lower (engine) rails only, then these members have to work as beams in differential bending to carry the torque T into the passenger compartment. This situation can effectively be present if there is a flexible or non-existent inner fender shear panel (e.g. with very large holes for drive shafts, air conditioning units, etc.).

Support for bending stiffness in the lower rail can be accomplished by incorporating this member into a grillage structure HH KK NN in the floor at the front of the compartment.

In the vehicle torsion load case, torque T is applied as couple $R_F S_F$ on the rails. The rails are supported at points H and N by the engine bulkhead and the floor cross-member KK respectively.

For force equilibrium

$$R_\mathrm{F} + P_4 - P_1 = 0$$

For moment equilibrium on the rail

$$R_\mathrm{F} L_1 = P_4 L_2 \quad \text{(taking moments about H)}$$

$$\therefore \qquad P_4 = R_\mathrm{F}\frac{L_1}{L_2} = \frac{L_1}{S_\mathrm{F} L_2}T \quad \left(\text{since } R_\mathrm{F} = \frac{T}{S_\mathrm{F}}\right) \qquad (6.1)$$

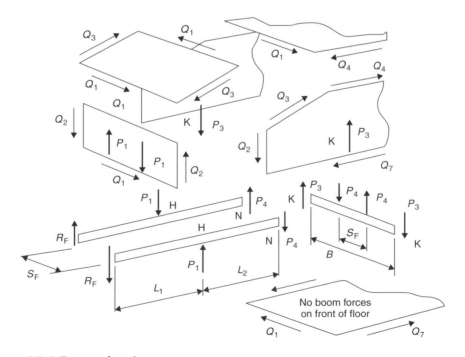

Figure 6.5 Grillage type frontal structure.

Thus:

$$P_1 = R_F \left(1 + \frac{L_1}{L_2}\right) = \left(1 + \frac{L_1}{L_2}\right) \frac{T}{S_F} \tag{6.2}$$

Note that P_1 on the bulkhead is bigger than R_F, since it has to react both R_F and P_4. The front parcel shelf is no longer important for introducing the torque loads to the compartment. With the longitudinal reaction Q_{x1} on the mid A-pillar from the parcel shelf absent, there is less local bending in the lower A-pillar. However, the rocker now experiences high local bending moments around point K. The floor no longer experiences longitudinal boom force reactions.

The floor cross-member also experiences high local bending moment at point N (Fig. 6.6). For the floor cross-member:

$$P_4 S_F = P_3 B \quad \text{(moment equilibrium)}$$

Hence

$$P_3 = P_4 \frac{S_F}{B} = \frac{L_1}{L_2 B} T \tag{6.3}$$

The distribution of edge forces Q in the compartment remains the same as in the baseline standard sedan torsion case (Chapter 5, section 5.3). So do the equilibrium equations for all of the compartment surfaces except for those of the engine bulkhead, the floor and the sideframe (see equations (5.21), (5.26) and (5.27) in Chapter 5). Assuming standard sedan type *rear* structure, these become:

$$Q_1 h_1 + Q_2 B = P_1 S_F \tag{5.21a}$$

$$-Q_1 L_5 + Q_7 B = -P_{RT} S_R \tag{5.26a}$$

$$-Q_2 r_2 + Q_3 r_3 + Q_4 r_4 + Q_5 r_5 - Q_6 r_6 + Q_7 r_7 = Q_{x2}(h_2 - Z) - P_3(X - L_2) \tag{5.27a}$$

using the notation of Figures 5.12. and 6.5.

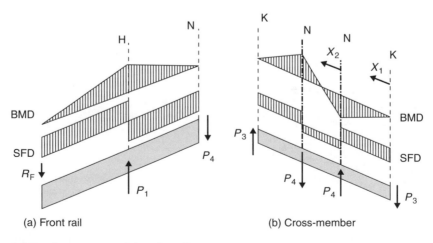

(a) Front rail   (b) Cross-member

Figure 6.6 Bending moment and shear force diagrams.

6.2.4   Grillage type frontal structure with torque tubes (Figure 6.7)

In the grillage type suspension support structure, the torque T on the vehicle, in the vehicle torsion load case, is carried in differential bending by the two engine rails. The underfloor grillage in the last example served only to provide a support moment for the engine rail at H.

An alternative to this is to provide torque tubes A H and H$'$ A$'$ from the bottoms of the A-pillars, to which the engine rails are attached at H and H$'$, as in Figure 6.7.

The vertical support reaction $P_1 = \pm R_F$ at H is carried as an in-plane load by the bulkhead. The bending moment $M_1 = \pm R_F L_1$ in the engine rail at H is carried by the torque tube, which consequently must be stiff in torsion.

As shown, the torque tube transmits torque between the two engine rails out to the lower A-pillar/rocker joint. This will cause very large local bending moments in the A-pillar, with a large step or reversal in the bending moment diagram at the torque tube attachment point.

A variation of this arrangement is to run the torque tube all the way across the front of the vehicle floor – AHH$'$A$'$. In the vehicle pure torsion load case, the moments M_1 on the engine rails at H and H$'$ will partly be reacted by a torque M_2 in member H–H$'$ and the remainder M_3 will pass as a torque out to the sideframes along members H–A and H$'$–A$'$ (Figure 6.8(a)). It is thus a statically indeterminate arrangement, so that the torque share between H–H$'$ and H–A (and H$'$A$'$) will depend on their relative stiffnesses. In the vehicle bending case, because of symmetry, all of the moment M_1 will pass out to the sideframe (Figure 6.8(b)).

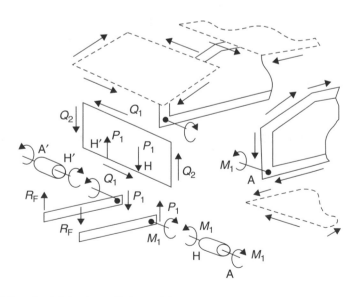

Figure 6.7 Grillage frontal structure with torque tubes.

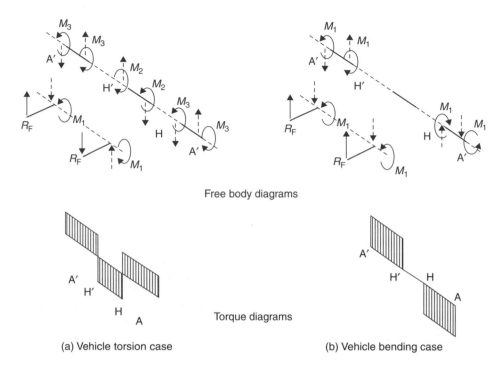

Free body diagrams

(a) Vehicle torsion case   Torque diagrams   (b) Vehicle bending case

Figure 6.8 Continuous torque tube loads.

6.2.5 Missing or flexible shear web in inner fender (see Figure 6.9)

If the shear web of the standard sedan front inner fender structure (Figure 5.12, Chapter 5) is absent or flexibly connected to the upper and lower boom members then the latter will work as two independent cantilever beams. They will thus carry bending moments rather than axial forces (i.e. they are beams rather than booms). This arrangement is much more flexible than the shear panel/boom assembly on the baseline sedan.

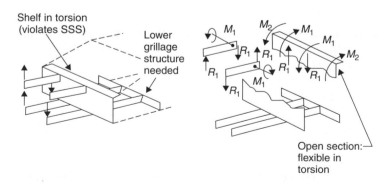

Figure 6.9 Flexible or missing inner fender shear panel.

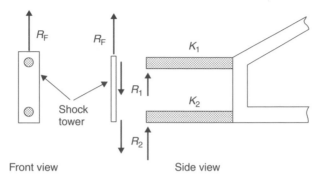

Front view   Side view

Figure 6.10 Upper and lower rails acting in parallel.

Furthermore, the high bending moments at the roots of these beams will load the parcel shelf in torsion and will load the floor with out-of-plane bending. This is also a very flexible arrangement.

The effect of this can be mitigated somewhat by designing the parcel shelf to carry torsion (closed section) and using a grillage arrangement in the floor to support the bending in the lower rail.

In this arrangement, if the suspension tower is reasonably stiff and is connected to both rails, then the upper and lower rails act as structures in parallel (Figure 6.10). This is a *statically indeterminate* arrangement and hence the load is shared between the rails in proportion to their stiffness in the vertical direction:

$$\text{i.e. } R_1/K_1 = R_2/K_2 \quad \text{or} \quad R_1/R_2 = K_1/K_2$$

where K_1 and K_2 are the vertical stiffnesses of the upper and lower rails.

6.2.6 Missing shear web in inner fender: upper rail direct to A-pillar

A way of by-passing parcel shelf torsion in the case of independent upper and lower rails is to position the upper rail so that it runs directly to the sideframe as shown in Figure 6.11.

The sideframe is suitably orientated to carry the reaction forces P_1 and moments $M_1 = P_1 L_1$. The moment reactions cause large bending moments in the A-pillar at waist level. Good joints, capable of carrying moments, are required at the junctions of the upper rails to the A-pillars.

It is instructive to consider the role of the suspension tower in this case. This is simplified as a triangular, planar simple structural surface in Figure 6.12.

If we adhere rigidly to the assumptions of the simple structural surfaces method, then torsion in the upper and lower longitudinal rails is ignored. In this case, consideration of the free body diagram (FBD) of the suspension tower reveals that this component is 'simply supported' by the rails. A moment balance will give the reaction forces R_1 and R_2 on the rails in terms of the suspension force R_F.

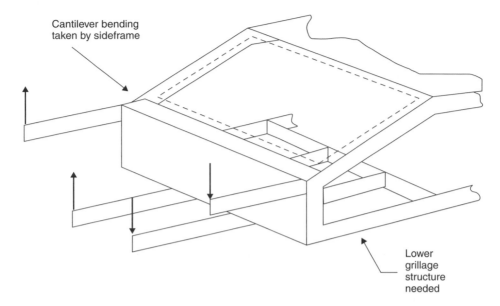

Figure 6.11 Upper rail attached directly to sideframe.

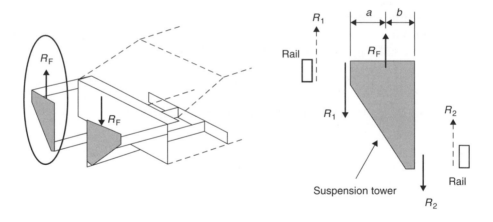

Figure 6.12 Suspension tower forces.

Taking moments about the lower rail connection:

$$R_1 = R_F b/(a + b) \qquad (6.4)$$

Vertical force equilibrium gives:

$$R_F = R_1 + R_2$$

Substituting from equation (6.4)

$$R_2 = R_F a/(a + b) \qquad (6.5)$$

Note that this is a statically *determinate* case. The values of R_1 and R_2 depend only on the position of the suspension strut force R_F. For example, in the extreme case that

$b = 0$ (i.e. force R_F acting in vertical plane through lower rail), then all of the load would be carried by the lower rail alone.

If, contrary to the assumptions of the simple structural surfaces method, the longitudinal rails *could* carry torsion, then the system would become statically indeterminate and rail and tower stiffness would then influence the load share between the rails.

6.2.7   Sloping inner fender (with shear panel)

This assembly has upper and lower rails offset laterally from each other, but with a shear panel between them. It represents a simplification of an inner fender system with an arbitrary-shaped shear panel. Lessons learnt from this will be extended to the arbitrary case in the next section.

The sloping inner fender arrangement (see Figure 6.13) has the advantage that the upper fender rail loads are fed straight into the body sideframe, by-passing the parcel shelf. It will be seen later, however, that the parcel shelf is still loaded. The presence of the shear panel makes this arrangement stiff.

The suspension load R_F is fed into the fender via a suspension tower, as shown in Figure 6.14. The tower is simplified to a triangular surface GIK as shown in the figure.

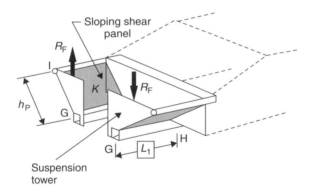

Figure 6.13 Sloping inner fender assembly, with effective shear panel.

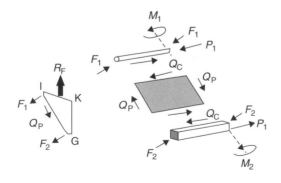

Figure 6.14 Free body diagrams of sloping fender components.

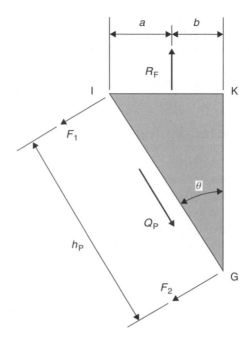

Figure 6.15 Free body diagram of suspension tower.

The tower experiences support force Q_P from the sloping shear web. Considering moments on the tower (see Figure 6.15), it may be seen that R_F causes a moment $R_F b$ about point G. Force Q_P cannot balance this, since its line of action is through G. This implies an out-of-plane force F_1 normal to the plane of the fender, acting at the upper rail attachment I, causing a balancing moment $F_1 h_P$ about point G. By similar reasoning there will be a similar force F_2 in the lower rail, normal to the fender.

Forces F_1 and F_2 will be carried by the upper and lower rails as out-of-plane transverse forces, causing bending in the rails about an axis which lies in the plane of the sloping fender (i.e. out-of-plane bending with respect to the boom-panel assembly).

This violates the simple structural surfaces method assumptions for 'boom-panel' assemblies. It is assumed here that the rails or 'booms' carry these out-of-plane forces because they have greater bending stiffness than the panel. It may safely be assumed that the in-plane shear force Q_P will be carried by the panel because of its high in-plane stiffness.

The forces may be calculated by applying equilibrium conditions as follows, using the notation in Figures 6.14 and 6.15.

Suspension tower

Resolving forces in the plane of the shear panel:

thus
$$Q_P = R_F \cos(\theta) \qquad (6.6)$$

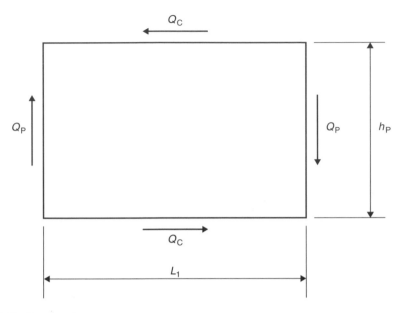

Figure 6.16 Shear panel.

Resolving forces normal to the shear panel:

$$F_1 + F_2 - R_F \sin(\theta) = 0$$

Take moments about point I:

$$F_2 h_P - R_F a = 0$$

$$F_2 = R_F a / h_P \qquad (6.7)$$

F_1 may be obtained by substituting in the above equations, or by taking

moments about point G: $\qquad F_1 h_P - R_F b = 0$

giving $\qquad\qquad\qquad\qquad\qquad F_1 = R_F b / h_P \qquad (6.8)$

Shear panel

For moment equilibrium:

$$Q_P L_1 - Q_C h_P = 0$$

Substituting from equation (6.6):

$$Q_C = Q_P (L_1 / h_P) = R_F (L_1 / h_P) \cos(\theta) \qquad (6.9)$$

Upper rail

For longitudinal force equilibrium (and substituting from equation (6.6)):

$$P_1 = Q_C = R_F (L_1 / h_P) \cos(\theta) \qquad (6.10)$$

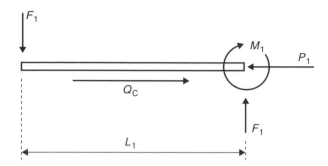

Figure 6.17 Upper rail forces in sloping inner fender.

Resolving forces in the rail normal to the panel:

$$F_1 = R_F b / h_P \quad \text{(see earlier)}$$

Taking moments about an axis in the plane of the panel (Fig. 6.17):

$$F_1 L_1 - M_1 = 0$$

Substituting from equation (6.8):

$$M_1 = F_1 L_1 = R_F (b / h_P) L_1 \tag{6.11}$$

Lower rail

By similar reasoning:

$$P_2 = Q_C = R_F (L_1 / h_P) \cos(\theta) \tag{6.12}$$

$$M_2 = F_2 L_1 = R_F (a / h_P) L_1 \tag{6.13}$$

Notes on sloping fender behaviour

F_1, F_2, M_1, M_2 are out-of-plane forces and hence violate simple structural surfaces method assumptions. The out-of-plane stiffness depends on the bending stiffness of the rails.

Since F_1 and F_2 (Fig. 6.14) are not equal, they can be thought of as resulting from the sum of two pairs of forces (see Figure 6.18):

(a) F_A causing local out-of-plane bending moment M; and
(b) F_B causing a local torsion couple $T_B = F_B h_P$ on the assembly.

F_A and F_B can be derived from F_1 and F_2 as follows:
Resolving forces in the rail normal to the panel:

$$2F_A \equiv F_1 + F_2 \tag{6.14}$$

Substituting from equations (6.7) and (6.8):

$$F_A \equiv (F_1 + F_2)/2 = R_F (a + b)/(2h_P) \tag{6.15}$$

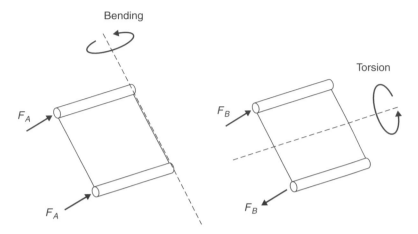

Figure 6.18 Pairs of forces on sloped assembly.

Taking moments about the torsion axis of the fender to get the local torque T_B applied to it:

$$T_B = F_B h_P \equiv F_1 h_P/2 - F_2 h_P/2$$

thus

$$F_B \equiv (F_1 - F_2)/2 = R_F(b - a)/(2h_P)$$

The forces F_A causing out-of-plane bending reflect the fact that the fender assembly is being subjected to bending about a non-principal axis, with a resulting lateral component of deflection. In this particular case, the principal axes are aligned with the plane of the shear panel. Force F_B, causing local torsion T_B, will be present if the load R_F is not applied through the shear centre of the assembly. In this case the shear centre is on the centreline of the web. This will be explained in the next section (6.2.8).

6.2.8 General case of fender with arbitrary-shaped panel

The problem of the principal axis directions and of the position of the shear centre have an effect in the case of fender-webs of general section. Readers unfamiliar with the background to these concepts are referred to Megson (1999).

In general the web sections are not symmetrical. The applied load R_F rarely aligns with the principal axis of the cross-section of the inner fender, so that lateral components of deflection are inevitable as indicated in Figure 6.19(a). If torsion is to be avoided, the section must be loaded through its shear centre. If the force R_F is offset from the shear centre by distance e (Figure 6.19(b)) then there is a net local torque of R_Fe on the assembly leading to twist. The position of the shear centre depends on the cross-sectional shape, and this is usually decided on factors other than structural ones. Similarly, the line of action of the suspension force is not under the control of the structural designer, so that it may be impossible to ensure that this coincides with the shear centre of the inner fender section.

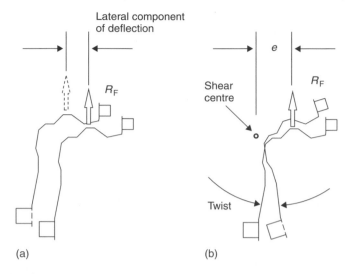

Figure 6.19 Effect of principal axis directions and shear centre position.

A common remedy is to include members, running across the vehicle between the front ends of the upper and lower rails. In the vehicle bending case, in which the loads are symmetrical, any tendency for twist or lateral deflection of the inner fenders can be balanced by tension or compression in these members. In the pure vehicle torsion case, the loads, and hence the distortions of the fender assemblies, are anti-symmetric so that more than simple tension or compression is needed in these cross-rails. A horizontal simple structural surface is required between the rails to resist the lateral deflection by its shear stiffness. The inner fender side rails and the added frontal cross-rails can form part of this if the corner joints are sufficiently stiff. An alternative is to include diagonal 'strut brace' tubular members from the tops of the suspension towers to an anchor point on the vehicle centreline at the parcel shelf/bulkhead junction.

6.3   Closed model variants

In this section we will consider closed model variants, that is different closed body configurations mounted on a common floor pan assembly. The standard sedan floor is often used for the estate car or for the hatchback. Sometimes the rear (trunk) floor is extended for the estate car or shortened for the hatchback but otherwise the rear floor structure remains little changed.

The bending load case can be treated in the same way as described in Chapter 5 and in more detail in Chapter 8. The rear door frame for both the hatchback and the estate car will function in a similar way to the rear panel of the saloon. That is, the sill of the rear door frame will carry bending and shear loads transferring forces into the sideframe.

For the torsion load case special care must be taken when analysing the loads in the rear door frame. A hatchback/estate car subject to the torsion load condition is shown

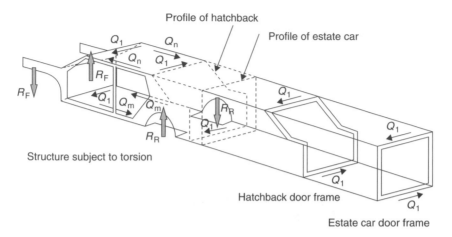

Figure 6.20 Torsion loading of hatchback/estate car.

in Figure 6.20. For clarity only the edge loads on the roof and the floor panels are shown.

6.3.1 Estate car/station wagon

The modern station wagon usually has a rear door that is near vertical and constructed in a single or near single plane. The rear door frame can then be represented by a single SSS and its primary structural function in the torsion case is to transfer shear from the roof panel to the floor. In Chapter 4, section 4.4 and Figure 4.8, we noted the importance of this SSS in providing this load path to maintain the torsional stiffness of the vehicle.

Figure 6.21(a) shows the loading on this rear door frame while the loads acting on a quarter of this frame are shown in Figure 6.21(b). The frame is loaded in complementary shear with the relationship between the loads given by:

$$Q_1 h - Q_2 w = 0$$

The shear forces are distributed uniformly along the members. The bending moments on the vertical member at the corner of $Q_1 h/4$ is of course equal to the bending moment on the top member $Q_2 w/4$. The loading on the other quarters of the frame is similar. It is important to note that the maximum bending occurs at the corner joints hence the design of these must be as stiff as possible and the sections of the frame must also be such that large deflections δw and δh as shown at Figure 6.21(c) do not occur. Each member of the frame has a point of contra/flexure as shown in Figure 6.21(c) and a constant shear force over the entire length.

For this model the rear suspension loads are applied to the sideframes. If the rear suspension loads are applied to longitudinal beams under the luggage floor additional loads may be transferred to the sill of the rear door frame. In which case these will of course then need to be included in the equilibrium equations.

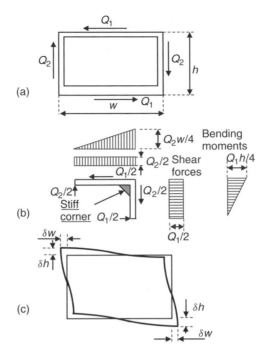

Figure 6.21 Estate car rear door frame.

6.3.2 Hatchback

In some countries the hatchback style of body is very popular and this has caused particular problems for the vehicle designer. Because of aerodynamic requirements to minimize drag, most designs are constructed with a sloping backlight with a short horizontal surface at waist height and then a vertical surface between waist and rear bumper. The rear door frame therefore is no longer a plane structure but has effectively three planes. Its structural function remains the same as for the rear door frame of the estate car, i.e. to transfer shear from roof to floor when the structure is subject to torsion.

If the structure is represented with four SSSs as shown in Figure 6.22 some interesting structural problems are discovered. The upper structure or U-shaped member above the waist can be considered as an SSS loaded in complementary shear where:

$$Q_1 a - Q_2 w = 0$$

Similarly the lower vertical U-shaped member is loaded in complementary shear:

$$Q_1 c - Q_3 w = 0$$

The short horizontal beams between these two will have the edge loads $Q_1/2$ acting at each end separated by the length b. The moment created could be balanced by a

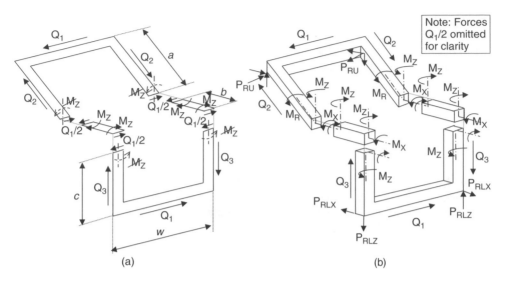

Figure 6.22 Rear door frame for hatchback.

moment at each end of M_z where:

$$2M_z - Q_1 b/2 = 0$$

At the rear end the M_z must then be reacted in the lower U-shaped member by an equal moment. This is unacceptable as the moment is applied normal to the SSS representing the lower member. At the front end the M_z is applied at an angle relative to the plane of the upper U-shaped member which again is unacceptable for an SSS. Therefore the rear frame is not satisfactorily modelled with these SSSs.

If the frame is modelled with closed sections as shown in Figure 6.22(b) the moments can be satisfactorily carried. If M_Z is applied at an angle to the plane of the upper member, there will be two components of moment, a torsion moment M_R in the sloping member and a torsion moment M_X in the horizontal member. As there is a second moment M_R applied at the other side, additional reactions P_{RU} acting against the sideframes are necessary to hold the upper frame in equilibrium. The forces P_{RU} act in the plane of the sideframe and normal to the plane of the sloping member. When analysing the sideframe they can be considered to have horizontal and vertical components:

$$P_{RU} w - 2M_R = 0$$

The short horizontal members will have a constant shear force $Q_1/2$ over the whole length with bending moments M_Z applied at each end giving contraflexure at mid-length.

The lower rear U-shaped structure will have the moment M_X applying bending to the vertical sides and M_Z applying torsion. Additional reactions P_{RLX} and P_{RLZ} acting against the sideframes will be necessary to maintain equilibrium for this member:

$$P_{RLX} w - 2M_Z = 0$$

$$P_{RLZ} w - 2M_X = 0$$

The edge loads Q_1, Q_2 and Q_3 are also applied to the closed section model shown at Figure 6.22(b).

The design implications of these loads are that all corner joints must be very stiff and capable of carrying bending moments, torsion moments and shear forces. In addition, the closed section members must be of adequate size to provide sufficient bending and torsional stiffness and strength. Great attention to detail design is therefore necessary.

6.3.3   Pick-up trucks

The pick-up truck is often a variant within a range of cars or light vans. There are several ways of achieving this, including the integral body approach, the use of 'through rails', and the separate chassis type. The design adopted will depend on many factors such as the format of the parent vehicle, production circumstances and the use to which the vehicle will be put.

(a) Separate chassis-frame type

For this type the body is mounted onto a completely separate ladder-frame chassis in an analogous arrangement to that of the pre-World War II car. Trucks of this type are usually purpose designed, rather than being derived from a car or light van range. Advantages of this approach include almost complete freedom for the mounting of body elements, and the ease of attachment of ancillary equipment such as winches etc. In production it is often economical to have all the mechanical parts mounted on a separate platform, to be 'mated' with the body later. This separation (in combination with elastomer body mounts) can also give a degree of acoustic isolation. There are also disadvantages. The ladder frame tends to be flexible both in bending and torsion, and the body-on-frame arrangement tends to be heavy. The mounting of a stiff body on a flexible chassis frame leads to 'fighting' between the two with consequent 'squeaking' and 'rattling' and the need for careful design of body mounts. This is discussed at length in Chapter 3, and was one of the original reasons for the adoption of integral structures in cars. Detailed discussion of the structural mechanics of separate frame vehicles is outside the remit of this book. Beerman's book (1989) is recommended for this subject.

(b) Van derived pick-up trucks using through rails

Some intermediate light vans (so-called 'panel vans') are of integral construction. These and some car-derived light vans contain 'through rail' underbody structures in which continuous rails run the full length of the vehicle under the floor, but which are welded to it so that they form an integral part of the body. The reason for this is for design variant adaptability. For example, it is easy to convert the design to a 'chassis cab' variant, in which the rear floor and structure above it are omitted. The protruding underfloor members are made into closed sections with closing plates, leaving a closed section rear chassis frame for mounting bodies/payloads of the customer's own design. This can form the basis of a pick-up truck. Some chassis cabs are supplied with the floor in place. The underfloor members have the additional advantage of providing support for the floor panel against the out-of-plane payload forces. Whether or not such a structure can be described as 'integral' depends on how well the cab can carry torsion loads, and on how well it is attached to the rails. If the cab does not contribute to the torsion stiffness then the structure reduces, in effect, to a 'separate chassis' type, in which case the underfloor grillage (including cross-members) must be designed with appropriate torsion capability, perhaps by using all closed section members.

(c) Pick-up trucks with integral structures

The true integral pick-up truck is more often seen where it has been created by the 'conversion' of a sedan type vehicle by replacing the body behind the B-pillar with a load platform, with substantial sidewalls, of typical pressed steel construction. An example is shown in Figure 6.23, along with a depiction of the structure of a light van in the same range.

(a)

(b)

Figure 6.23 (a) Pick-up truck; (b) Van structure (both courtesy of Volkswagen AG).

SSS idealization of a hypothetical integral pick-up truck

The structural layout of integral pick-up trucks from different manufacturers varies widely. An SSS idealization of a hypothetical integral pick-up truck is shown in Figure 6.24. At the front, this particular example has an underfloor grillage structure to carry the front suspension and engine loads. This consists of two lower front rails supported by the front and rear cab bulkheads, which then carry the support loads out to the cab sideframes.

The rear suspension loads are reacted by wheel-box suspension towers attached to the rear load platform sidewalls. In this example, the latter act as deep boom-panel cantilevers fixed to the lower B-pillars of the cab side. Since the load platform floor is loaded out-of-plane by the payload, it will require stiffening to enable it to carry these forces out to the sidewalls which support it. The stiffening could take the form of lateral corrugations (see Chapter 7) or of underfloor cross-members. The in-plane shear stiffness of the floor can provide lateral support to the sidewalls.

A simplified bending load case is shown in Figure 6.25. This includes only the engine weight, and the payload simplified to a single force. This case is not pursued in detail here, but an interesting feature emerges. The junction between the B-pillar and the load platform sidewalls is in a region of high bending moment. This moment is reacted by the couple formed by the boom forces K_{X2} in the sidewalls at the cab attachment point (see Figure 6.25). The reaction to the upper boom force causes a high local bending moment in the B-pillar at waist level. This may require enlargement of the B-pillar section.

For the torsion case in this vehicle, the cab acts as the 'torsion box'. The luggage platform sidewalls act as levers to enable the rear suspension forces R_R to apply a torque to this box. The simple structural surface edge loads are shown in Figure 6.26.

The directions of the edge loads on many of the panels are determined by the logic of complementary shear forces. On the front and rear cab bulkheads the edge force directions (for Q_3 and Q_4) are not so obvious because of the extra forces P_1 and P_3. If, in a simple structural surfaces idealization, the direction initially chosen for Q_3 and Q_4 is incorrect then these forces will have negative values in the results. Using the method of section 6.2.3 and the notation in Figure 6.26, the reaction forces P_1 and P_3 on the lower front rails are:

$$P_1 = R_F(L_1 + L_2)/L_2 = T(L_1 + L_2)/(S_F L_2)$$

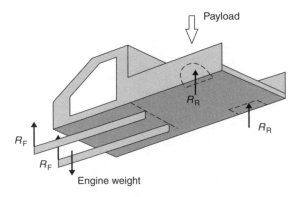

Figure 6.24 Hypothetical integral pick-up truck: SSS idealization.

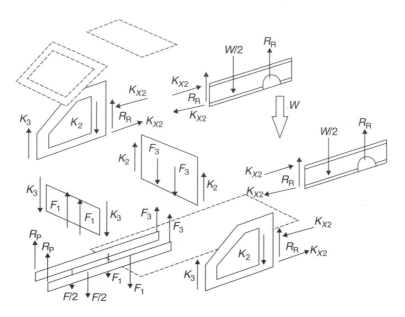

Figure 6.25 Simplified bending load case on an integral pick-up truck.

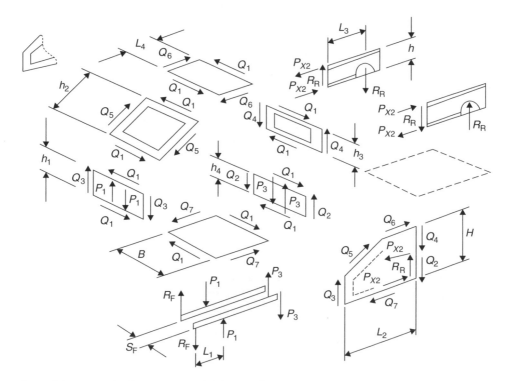

Figure 6.26 Torsion load case on integral pick-up truck.

and $$P_3 = R_FL_1/L_2 = TL_1/(S_FL_2)$$

where T is the torque applied to the vehicle.

Note that, as mentioned in section 6.2.3, the force P_1 on the front bulkhead is greater than the suspension force R_F. If the rear ends of the lower rails are mounted on a cross-member further forward than the rear bulkhead, then this force (P_1) will be bigger still. Moment equilibrium on the load platform sidewalls gives the values of the boom forces P_{X2} as:

$$P_{X2} = R_RL_3/h = TL_3/(Bh)$$

The equilibrium equations for the cab torsion box surfaces are shown below.

$$Q_1h_2 - Q_5B = 0 \text{ windshield}$$
$$Q_1L_4 - Q_6B = 0 \text{ roof}$$
$$Q_1h_3 - Q_4B = 0 \text{ backlight}$$
$$Q_1L_2 - Q_7B = 0 \text{ floor}$$
$$Q_1h_1 - Q_3B = +P_1S_F \text{ front bulkhead}$$
$$Q_1h_4 - Q_2B = +P_3S_F \text{ rear bulkhead}$$
$$Q_2L_2 + Q_4L_2 + Q_6(H - h_1) + Q_7h_1 = R_RL_2 + P_{X2}h_1 + P_{X2}(h_4 - h_1) \text{ cab side}$$

(The equation for the cab side comes from taking moments about the junction between the upper and lower A-pillars.)

These seven simultaneous equations can be solved for the seven unknown edge forces Q_1 to Q_7. As before (Chapter 5, section 5.3) this is most easily done by setting the equations in matrix form and using a standard solution method. Most spreadsheet computer programs have this equation solving capability.

Notes on integral open pick-up trucks

1. B-pillar loading

For the vehicle torsion case, as in the vehicle bending load case, the upper boom forces P_{X2} (Figure 6.26) in the load platform sidewalls are reacted by the B-pillar at waist level. As before, this will tend to give a high local bending moment in the B-pillar at this point, as shown in Figure 6.27(a). However, the B-pillar also forms part of the ring frame around the door opening. The edge shear forces Q_2 to Q_7 on this ring frame (Figure 6.26) lead to local contraflexure bending in the members around its perimeter, including the B-pillar as described in Chapter 5. In the B-pillar, this combines with the bending moment caused by the reactions P_{X2} at waist level to give a resulting bending moment distribution as shown in Figure 6.27(b). The B-pillar will require a large cross-section to carry these bending moments

2. 'Faux' integral open pick-up truck

In some integral open pick-up trucks, or similar integral open sports vehicles, there is no rear bulkhead to the cab, giving free access between the cab and the open rear platform. Clearly, in the torsion case for such vehicles, the cab will no longer be a 'closed box' structure, so that the system of edge shear forces described above (see Figure 6.26) will break down. In effect this gives a 'faux pick-up truck' structure, in

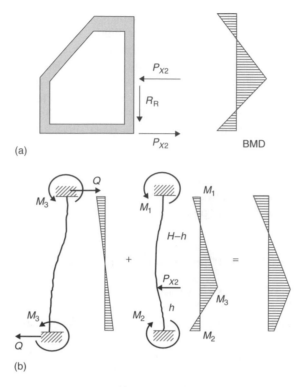

(a)

BMD

(b)

Figure 6.27 (a) Bending moment due to reaction P_{X2}; (b) combined bending moment distribution in the B-pillar.

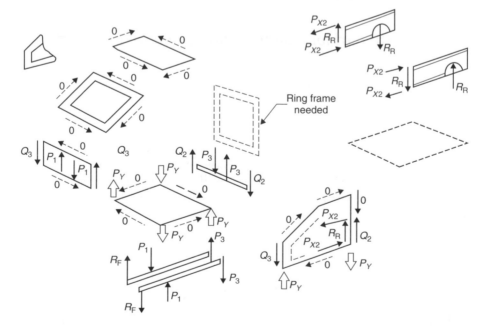

Figure 6.28 Faux open pick-up vehicle structure.

which torsion is carried by relatively weight-inefficient out-of-plane load paths in the dash structure or in the underbody platform (see Figure 6.28).

One possible way of overcoming this and restoring the closed box structure to the cab is to position a ring beam around the opening, as shown in Figure 6.28. This can incorporate some of the members which are already present, such as the B-pillars, with new roof and floor cross-members. In this example, a floor cross-member would be needed anyway to react forces P_3 (or F_3 in the vehicle bending case) from the front longitudinal members. The frame must be able to resist shear in its own plane, so that the ring frame edge members must be continuous around the perimeter, and they must be stiff in bending locally about an axis normal to the plane of the frame. Since local bending moments are maximum at the corners, it is essential that the frame corner joints are stiff in bending. In effect, this solution is similar to that described for the 'faux sedan' in Chapter 5 (see Figure 5.19).

6.4   Open (convertible/cabriolet) variants

Vehicles without a roof, such as convertibles/cabriolets, or in which the roof plays no structural part, will be referred to here as 'open vehicles'. The absence of a roof or upper sideframe means that the passenger compartment of the open vehicle will *not* consist of a closed box of simple structural surfaces in shear.

This has a profound effect on the load paths in the structure for both the bending and torsion load cases. For the torsion load case in particular, extra measures must be taken to ensure satisfactory performance. These will almost inevitably mean that the convertible will be heavier than a standard integral sedan (providing the latter has a satisfactory 'closed' passenger compartment).

6.4.1   Illustration of load paths in open vehicle: introduction

The load paths in the convertible will be illustrated for the vehicle shown in Figure 6.29, with front and rear end structures similar to the 'standard sedan'. Deep inner fenders front and rear are supported in shear by the compartment bulkhead while the top and bottom flange (or 'boom') forces are carried by parcel shelves and floor, respectively.

Inspection of the connection of the parcel shelf to the A-pillar reveals the absence of the upper sideframe to act as an open truss. The lower A-pillar (Figure 6.29) carries all the parcel shelf reaction, and hence needs to be stiff in bending and shear. 'Fill-in' shear panels A and C with appropriate flanges carry the shear and bending down from the parcel shelves to the compartment sidewall (or rocker panel). This is necessary for both torsion and bending load cases. The lower sidewall works alone in bending, and so must also be stiff, but it tends to be less stiff than the sideframe on a sedan car because it is less deep.

6.4.2   Open vehicle: bending load case

The simple structural surface edge loads for an open vehicle in a bending case are shown in Figure 6.30. For the sake of simplicity, a single vertical load P_Z is distributed

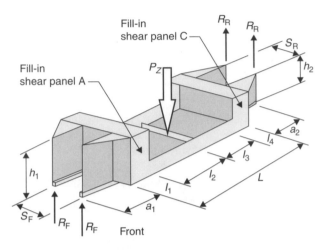

Figure 6.29 Simple structural surfaces model of open vehicle.

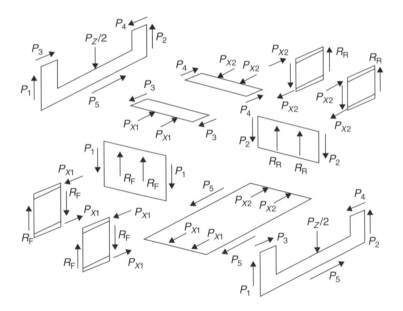

Figure 6.30 Simplified bending load case for open vehicle.

by a cross-member equally to the two sidewalls (as forces $P_Z/2$). This is a symmetrical load case, so the forces on each side are identical.

The front and rear ends in this vehicle behave in exactly the same way as the standard sedan as discussed in Chapter 5, section 5.2, so the analysis will not be repeated here. For this vehicle the rear fenders, parcel shelf and bulkhead behave identically to those at the front, but the magnitudes of the edge forces may be different because of different sizes, a_2, h_2, etc., and different front and rear wheel reactions. Using the notation in

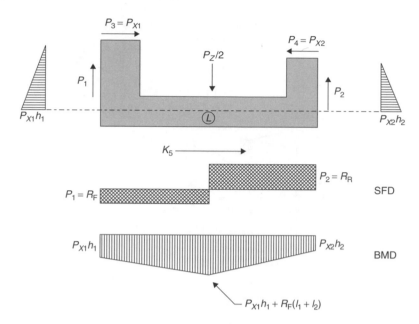

Figure 6.31 Sidewall shear force and bending moment diagrams for simplified bending load case.

Figures 6.29 and 6.30 the edge forces in the end structures may be summarized as:

	Front	*Rear*
Upper and lower boom force	$P_{X1} = R_{\mathrm{F}}a_1/h_1$	$P_{X2} = R_{\mathrm{R}}a_2/h_2$
Bulkhead side reaction	$P_1 = R_{\mathrm{F}}$	$P_2 = R_{\mathrm{R}}$
Parcel shelf to pillar force	$P_3 = P_{X1} = R_{\mathrm{F}}a_1/h_1$	$P_4 = P_{X2} = R_{\mathrm{R}}a_2/h_2$

The floor is in tension, as might be expected in the bending case. Floor edge force P_5 (same on both sides) is present only if P_{X1} and P_{X2} are of different magnitude, so that $P_5 = P_{X2} - P_{X1}$. If this is the case, then the outer sections of the floor are in shear between the lines of action of P_{X1} and P_5.

The sidewalls are the major members carrying bending along the passenger compartment. They need to be stiff in bending because there is no upper sideframe to act as a truss. At the ends, bending moments are fed into the sidewall from forces P_3 and P_4 via the A- and C-pillar fill-in panels and their flanges. The overall shear force and bending moment diagrams for the sidewall are shown in Figure 6.31.

6.4.3 Open vehicle: torsion load case

The open car suffers in the torsion case from not having a 'closed box' passenger compartment.

The fender/bulkhead/parcel shelf structures behave in much the same way as in the closed standard sedan. Thus, referring to Figures 6.32 and 6.29, the equilibrium equations are:

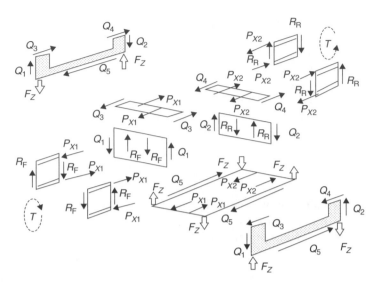

Figure 6.32 Edge loads in torsion load case on open vehicle.

	Front	Rear
Suspension force	$R_F = T/S_F$	$R_R = T/S_R$
Upper and lower boom force	$P_{X1} = R_F a_1/h_1$	$P_{X2} = R_R a_2/h_2$
Bulkhead side reaction	$Q_1 = R_F S_F/B = T/B$	$Q_2 = R_R S_R/B = T/B$
Parcel shelf to pillar force	$Q_3 = P_{X1} S_F/B$	$Q_4 = P_{X2} S_R/B$
Floor (in-plane)	$P_{X1} S_F + P_{X2} S_R - Q_5 B = 0$	

The main difference comes when considering the sidewall. Close inspection of Figures 6.32 and 6.33 reveals that all the edge forces on the sidewall create *moments in the same direction*. There are thus no complementary balancing moments. The only way this out-of-balance moment can be neutralized is for couples from *out-of-plane* forces F_Z from the *floor*.

Thus, the simple structural surfaces 'in-plane shear panel' assumptions have broken down, and the surfaces must rely on their out-of-plane stiffness. This type of structure tends to be highly flexible in torsion unless remedial measures are taken. Twisting tends to centre on the engine fire wall, with the sidewalls acting as 'levers' applying torsion from the rear suspension forces.

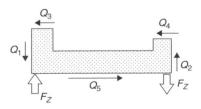

Figure 6.33 Open vehicle sidewall in torsion load case.

6.4.4 Torsion stiffening measures for open car structures

Many different approaches have been tried over the years to improve the torsion stiffness of open vehicles, with varying degrees of success. It is important, not only to achieve increased stiffness, but to do so without an excessive weight penalty.

The particular approach used will depend on the type of open vehicle under consideration. Stiffening measures for purpose-built open sports cars are likely to be different from those used in 'cabriolet conversions' of mass produced integral sedans.

Two widely used methods are:

(a) Adding or building in a torsionally stiff grillage type structure into the vehicle floor.
(b) Finding areas within the vehicle body which can be 'boxed in' to give torsionally stiff elements. It is then necessary to provide adequate structural connections from the torsionally stiff box to the rest of the body.

Since many open vehicles have a 'sporty' aspect, an additional approach is to add a 'roll-bar' or a 'T-roof' to the vehicle to tie the two sides together in torsion. In a sense, this is an attempt to replace part of the roof structure removed in the first place. It also amounts to adding cross-members to the vehicle, and this is another approach which has been used also. Such methods are usually not very effective on their own, but they are often incorporated as subsidiary methods to be used in association with the major approaches listed above.

The torsional stiffening approaches outlined in (a) and (b) will now be discussed with examples.

(a) Torsionally stiff grillage in floor

Several different versions of this are available, they include:

• Cruciform braced members in the floor. This was discussed in Chapter 3 (see Figures 3.10 and 3.11 and associated text). It has the advantage that the members used in the brace do not themselves need to be torsionally stiff, so that open sections can be used.
• Torsionally stiff tubular members in the floor. These may take the form of longitudinal 'backbone' members, or sometimes they may be laterally positioned.
• A full 'grillage' of tubular interconnected members, each with good local torsion and bending stiffness and with joints which are good at transmitting torsion and bending.

Tubular ' backbone' members This approach is often used in specialist sports cars, conceived from the outset as open vehicles. They often consist of a separate longitudinal 'backbone chassis frame' with added bodywork. These could be termed 'longitudinally aligned torsion boxes'.

Torsion stiffness for this comes from a large cross-section area of the tube. The torsion constant J for this is proportional to the enclosed area squared (A_E^2), so a large tube ($\sim 300 \times 200\,\text{mm}$) gives a certain amount of stiffening. It is essential that the backbone be of *closed* section, since open sections are very flexible in torsion.

Two examples: the Lotus sheet steel backbone structure, and the TVR backbone, made of bays of triangulated tubes, were shown in Chapter 3 (Figures 3.14 and 3.15).

Another good example of this approach is the Chevrolet Corvette (Figure 6.34). The structure of this consists of a closed central tube which is structurally tied to long hydroformed outboard longitudinals, so that it is a mixture of the 'backbone' and 'punt' types. Note that all of the examples in this section are specialist sports cars with separate frames.

Full underfloor grillage structure This is an approach more commonly used in 'cabriolet conversions' of integral body vehicles, because it is relatively easy to incorporate existing closed members (such as the rocker) into the required grillage.

An example of this is shown in Figure 6.35. The integral structure of a Japanese 'micro-car' was turned into a cabriolet by removing the roof. To restore torsional

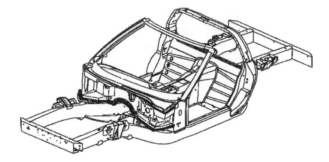

Figure 6.34 Chevrolet corvette structure (courtesy of General Motors Corporation).

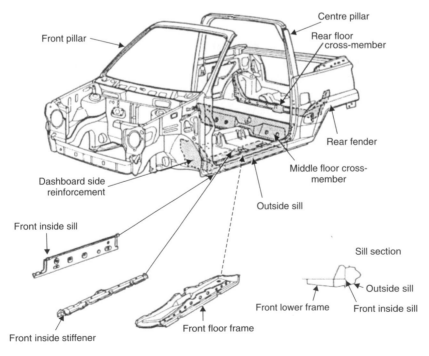

Figure 6.35 City car cabriolet conversion (courtesy Honda UK Ltd and *Automotive Engineering* magazine).

stiffness inevitably lost by removing the roof, the floor of the vehicle was converted into a grillage. The longitudinal rockers were turned into multi-cell box members by the addition of additional sheet steel closing plates. The enclosed area A_E of these members was thereby increased by 3.5 times. Good torsional connection to the dash was achieved by 'butterfly' ends on the fronts of the enlarged rockers, welded to the dash. A new closed section middle floor cross-member was also created by adding closing plates to the seat cross-member. A roll-hoop, and an additional rear cross-member were also incorporated.

It was reported (*Automotive Engineering* magazine 1984) that the cabriolet had the same torsion stiffness as the sedan, with the cabriolet body-in-white weighing 60 kg more than that of the closed vehicle. Such weight increases are by no means unusual in 'cabriolet conversions'. They simply underline the difficulty in achieving high torsion stiffness in open vehicles.

(b) Addition of a 'closed-simple structural surfaces-box', using space and components which are there for other purposes (see Figures 6.37 and 6.38)

Such boxed-in regions usually run across the car, tying the two sides together torsion-ally. The sidewalls act as 'levers' applying torsion to the box. As such, these could be called 'laterally aligned torsion boxes'. This method is also applicable to longitudinally aligned boxes if an appropriately shaped structure is available. Boxes in either direction are capable of providing torsion stiffness, as shown in Figure 6.36.

Typical areas used for such boxes are:

(i) 'Boxing in' the luggage compartment or the engine compartment. This is done by ensuring that all openings in the region are surrounded by good, complete ring frames with stiff corner joints, and by ensuring structural continuity of the other sides of the box. This will be analysed in section 6.4.5, see Figure 6.38.

(ii) In the region of the engine bulkhead, parcel shelf and (wide) lower A-pillars. An additional ring frame is added, consisting of the instrument panel, wider lower A-pillar, and a floor cross-member (possibly *under* the floor). It is essential that all four corners of this frame have good joints and that the frame is well connected to the rest of the box. A diagram of the edge loads in such an arrangement is given in Figure 6.37 but the case is not analysed here.

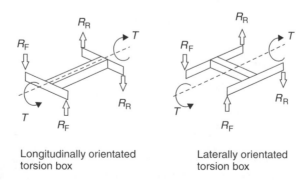

Longitudinally orientated torsion box

Laterally orientated torsion box

Figure 6.36 Longitudinally and laterally orientated torsion box.

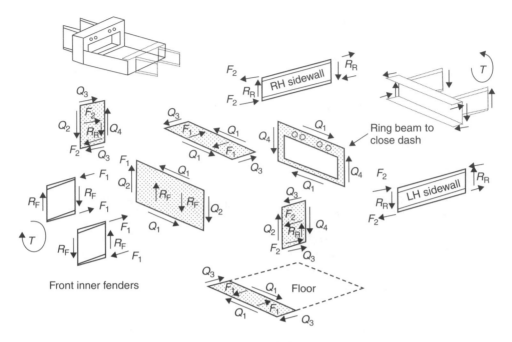

Figure 6.37 Example of open vehicle with torsion box in dash region.

(iii) In the region under the rear seat or near the fuel tank (there is often a step in the floor at this point, with a certain amount of available space).

6.4.5 Simple structural surfaces analysis of an open car structure torsionally stiffened by 'boxing in' the engine compartment

To simplify the calculation, the vehicle analysed will have a simple single cross-member rear suspension mounting.

The 'box' consists of (Figure 6.38):

(i) the inner front fenders (upon which the suspension loads R_F are applied, forming torsion couple T);

(ii) part of the passenger compartment front bulkhead (shaded in the figure);

(iii) a 'radiator surround' ring beam;

(iv) an upper ring consisting of the fender flanges, the parcel shelf and a hood-slam cross-member at the front, again good corner joints are needed;

(v) a lower ring incorporating the engine rails, the main floor and a front lower cross-member.

Note that the 'box' occupies only a small part of the width of the vehicle. Torque from this box is transmitted to the sidewalls by differential bending in the parcel shelf and the floor, so that edge forces Q_{X1} and Q_F form a couple which resists the bending

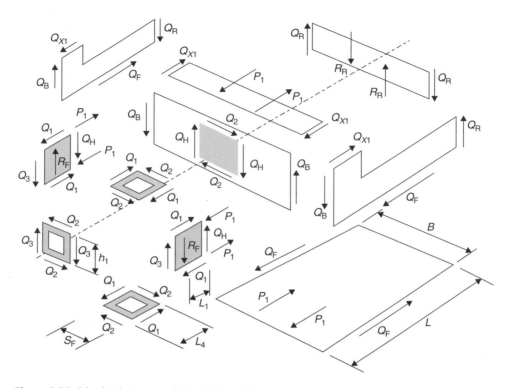

Figure 6.38 Edge loads in open vehicle with 'boxed' engine compartment.

moment at the front of the sidewall. The two sidewalls act as 'levers' transmitting the torque to the rear suspension mounts.

The distribution of edge loads is as shown in Figure 6.38. The moment equilibrium equations are as follows. It is difficult to assign directions to some of the forces (e.g. Q_H) since these depend on the positions of other forces (e.g. R_F). If we choose wrongly before the calculations, the value of the force will come out negative in the results.)

Radiator frame
$$Q_2h_1 - Q_3S_F = 0 \tag{6.16}$$

Engine compartment top and bottom
$$Q_1S_F - Q_2L_4 = 0 \tag{6.17}$$

Fender left hand and right hand

Vertical force equilibrium $Q_3 + Q_H = R_F$ (6.18)

Moment equilibrium (take moments about R_F)

$$Q_1h_1 + Q_3(L_4 - L_1) - Q_HL_1 - P_1h_1 = 0 \tag{6.19}$$

Front bulkhead (passenger compartment)
$$Q_2h_1 + Q_HS_F - Q_BB = 0 \tag{6.20}$$

Parcel shelf

$$P_1 S_F - Q_{X1} B = 0 \tag{6.21}$$

Sidewall

Force equilibrium $\qquad Q_{X1} - Q_F = 0 \tag{6.22}$

Moment equilibrium (take moments about bottom front corner)

$$Q_{X1} h_1 = Q_R L \tag{6.23}$$

Rear bulkhead

$$Q_R B = R_R S_R \qquad \therefore \qquad Q_R = T/B \quad (T = \text{torque applied to vehicle})$$

Rear suspension

$$R_R = T/S_R$$

Front suspension

$$R_F = T/S_F$$

The above equations can be put in matrix form as shown below. As shown they have been included in an order which ensures non-negative terms on the 'leading diagonal' (e.g. coefficient 5 in the fifth equation etc.). This enables solution by a simple computer program based on Gauss elimination without partial pivoting. 'Spreadsheet' computer programs can do this.

Variable $\quad Q_1 \quad Q_2 \qquad Q_3 \qquad\quad Q_H \quad P_1 \quad Q_B \quad Q_{X1} \quad Q_F$
Equation

	Q_1	Q_2	Q_3	Q_H	P_1	Q_B	Q_{X1}	Q_F			
(6.17)	S_F	$-L_4$	0	0	0	0	0	0	Q_1		0
(6.16)	0	h_1	$-S_F$	0	0	0	0	0	Q_2		0
(6.19)	h_1	0	$(L_4 - L_1)$	$-L_1$	$-h_1$	0	0	0	Q_3		0
(6.18)	0	0	1	1	0	0	0	0	Q_H	$=$	R_F
(6.21)	0	0	0	0	S_F	0	$-B$	0	P_1		0
(6.20)	0	h_1	0	S_F	0	$-B$	0	0	Q_B		0
(6.23)	0	0	0	0	0	0	h_1	0	Q_{X1}		$Q_R L$
(6.22)	0	0	0	0	0	0	1	-1	Q_F		0

These equations were tried for a vehicle with:

$$S_F = 1040\,\text{mm} \qquad h_1 = 640\,\text{mm}$$

$$L_1 = 480\,\text{mm} \qquad L_4 = 1080\,\text{mm}$$

$$B = 1360\,\text{mm} \qquad L = 1620\,\text{mm}$$

and for a unit torque (i.e. $T = 1$) this gave:

$Q_1 = 2.31 \times 10^{-3}$ N giving shear flow $q_1 = 2.14 \times 10^{-6}$ N/mm

$Q_2 = 2.23 \times 10^{-3}$ N $q_2 = 2.14 \times 10^{-6}$ N/mm

$Q_3 = 1.37 \times 10^{-3}$ N $q_3 = 2.14 \times 10^{-6}$ N/mm

$Q_H = -4.09 \times 10^{-4}$ N $q_4 = -6.38 \times 10^{-6}$ N/mm

$P_1 = 3.90 \times 10^{-3}$ N

$Q_B = 7.35 \times 10^{-4}$ N $q_B = 1.15 \times 10^{-6}$

$Q_{X1} = 2.98 \times 10^{-3}$ N

$Q_F = 2.98 \times 10^{-3}$ N $q_F = 1.84 \times 10^{-6}$

Notes

(i) The negative value for Q_H means that we chose the wrong direction for the force in Figure 6.38.

(ii) Note that q_1, q_2 and q_3 are equal as expected, since they are complementary shear flows on the 'radiator surround' and on the top and bottom surfaces of the engine compartment. Q_1 is the *net* force on the top and bottom edges of the inner fender surfaces, so that q_1 is the *average* shear flow on these edges.

(iii) The actual edge forces and shear flows could be obtained by multiplying the above results (for a *unit torque* $T = 1$ Nmm) by the actual torque to be applied.

7

Structural surfaces and floor grillages

7.1 Introduction

Integral car bodies are three-dimensional structures largely composed of approximately planar subassemblies, which we have called simple structural surfaces.

The methods of Chapters 4, 5 and 6 may be used to evolve a good basic layout for the body structure, incorporating continuous load paths, and to determine the edge loads passing globally between the main body surfaces. The next step is to ensure that the individual structural surfaces themselves can carry the loads imposed on them in a weight efficient way. Two things are needed for this. First, there must be an understanding of how different types of structural surface work, to avoid local structural discontinuities and to ensure the choice of the lightest type of structural surface for a given application. Second, approximate methods for calculating internal member loads within surfaces are required to enable initial sizing.

Planar structural subassemblies in integral car bodies may be divided into two groups based on their functions:

(a) Those which carry in-plane loads, referred to in this book as simple structural surfaces. Examples are the roof panel and the sideframe assembly.
(b) Those carrying out-of-plane loads. The best examples of these are passenger compartment floors. These are basically grillage structures.

Some surfaces, such as the floor, carry both types of load but the two functions can be treated separately. Thus, when the floor is acting as a grillage (to support seat and payload forces), the required out-of-plane stiffness comes from attached beam members. The actual floor panel, being very thin, makes a negligible contribution in this case. Conversely, when the floor is acting as part of the 'torsion box' in the passenger compartment, the requirement is for high shear stiffness. In this case, since the floor panel is much stiffer in-plane than the attached beams, it carries almost all of the shear load and the effect of the attached beams is negligible.

This chapter deals first with structural surfaces loaded in-plane, including sideframe structures, and then with floor grillage structures.

7.2   In-plane loads and simple structural surfaces

Most of the global loads on an integral car body are carried internally by planar structures which are stiff in their own plane. These were defined in Chapter 4 as 'simple structural surfaces'.

Such structures tend to be very stiff for in-plane loads such as in-plane shear forces Q, in-plane compression or tension P_X, P_Y and 'in-plane' bending (i.e. M_Z about the axis normal to the plane), as shown in Figure 7.1. They are flexible for out-of-plane loads such as out-of-plane shear forces P_Z, twist (torsion) $T_X T_Y$ about in-plane axes and bending $M_X M_Y$ about in-plane axes. It is assumed in the simple structural surfaces method that these elements are so much stiffer in-plane that the out-of-plane stiffness is negligible and can be ignored.

Structures which are flexible in-plane are *not* considered as simple structural surfaces, and will tend to cause unsatisfactory breaks in load paths in integral body structures. They include panels with very large, unreinforced cut-outs or missing edges, discontinuous panels or frames. Open ring frames with pin-jointed or flexible-jointed corners, or with very flexible sides and frames with missing edges also come into this category.

Typical examples of planar subassemblies which act as effective simple structural surfaces and thus confer satisfactory performance on the car body include:

(a) thin walled panels, including those with relatively small or stiffened cut-outs and panels with stiffeners;
(b) triangulated planar trusses;
(c) stiff jointed ring frames, for example windshield surrounds or body sideframes;
(d) combinations of the above, for example the sideframes of two-door cars.

Types (a), (b) and (c) will now be considered in turn. A comparison of the structural effectiveness of the different types, based on stiffness, will then be made.

7.2.1   Shear panels, and structures incorporating them

One role in which simple structural surfaces (and thin panels in particular) are highly effective is in carrying shear forces Q.

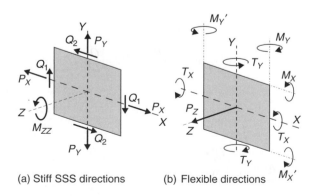

(a) Stiff SSS directions   (b) Flexible directions

Figure 7.1 Stiff and flexible directions for simple structural surfaces.

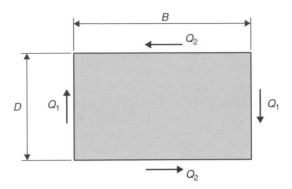

Figure 7.2 Panel in shear.

Considering equilibrium on the panel shown in Figure 7.2:

(i) For vertical and horizontal force equilibrium, the forces Q_1 on the left and right must be equal to each other, as must those, Q_2, on top and bottom.

(ii) The forces Q_1 are separated by distance B, and hence apply a couple $Q_1 B$ to the panel. For moment equilibrium, this must be balanced by an equal and opposite couple $Q_2 D$ ('complementary couple') caused by additional shear forces Q_2 top and bottom.

Thus: $$Q_1 B = Q_2 D \quad (Q_1 \text{ and } Q_2 \text{ are 'complementary shear forces')} \quad (7.1)$$

Shear flow and shear stress

The 'shear flow' q_1 (shear force per unit length) along the vertical edges is:

$$q_1 = \frac{Q_1}{D} \tag{7.2a}$$

The 'complementary shear flow' on the horizontal edges is:

$$q_2 = \frac{Q_2}{B} \tag{7.2b}$$

But $$Q_2 = Q_1 \frac{B}{D}$$

$\therefore$ $$q_2 = Q_1 \frac{B}{D} \times \frac{1}{B} = \frac{Q_1}{D}$$

i.e. $$q_1 = q_2 \tag{7.3}$$

Every shear flow q will always have an equal complementary shear flow normal to it.

The edge forces causing shear flow q will set up shear stress τ in the material along the perimeter. For shear panels, the shear stress is uniform across the thickness of the material (see Figure 7.3)

The average shear stress τ_{AV} along the panel edge is thus:

$$\tau_{AV} = Q_1/A_S = Q_1/Dt$$

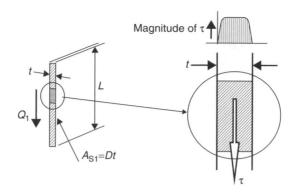

Figure 7.3 Shear stress on edge of panel.

where Q_1 = total shear force on edge of panel, A_S = shear area, D = length of panel edge, and t = panel thickness.

Note that average shear flow $q_1 = Q_1/D$, so that the relation between shear flow q_1 and shear stress τ_{Av} is:

$$q_1 = t\tau_1 \tag{7.4}$$

Like shear flow, shear stresses τ always come in complementary pairs.

Stiffness of shear panel

An approximate shear stiffness K_{SHEAR} of a shear panel may be calculated from the classical shear stress strain rule (Figure 7.4):

$$\text{shear stress} = \text{shear modulus} \times \text{shear strain}$$

$$\tau = G\gamma$$

Assuming constant shear stress and shear strain across the panel:

$$\tau = \frac{Q_1}{tD} = \frac{\text{Shear force}}{\text{Shear area}}$$

$$\gamma = \frac{\Delta}{B} \quad (\Delta = \text{Shear deflection})$$

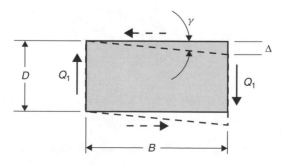

Figure 7.4 Shear deflection.

∴

$$\frac{Q_1}{tD} = G\frac{\Delta}{B}$$

$$Q_1 = K\Delta = \frac{GtD}{B}\Delta$$

i.e.   the shear stiffness of the panel $= K = GtD/B$ (7.5)

The volume (related to weight) of the *shear panel* is:

$$VOL_{\text{SHEAR PANEL}} = DBt$$

Shear panel as part of an assembly

The edge shear forces Q_1 and Q_2 will have to be reacted by adjacent structural elements. These may be (a) other shear panels (or other simple structural surfaces), or (b) edge 'booms' or flanges.

(a) Shear panel assemblies

The role of shear panels in car passenger compartment 'torsion boxes' has already been discussed in detail in Chapters 4 and 5. As an example, consider a floor with a longitudinal 'transmission tunnel' (or longitudinal 'services tunnel' in the case of a front wheel drive vehicle) subject to shear force Q_1 (see Figure 7.5):

Shear force Q_1 can be carried across the floor, being passed up and over the tunnel as shown, as long as there is sufficient attached structure to react the complementary shear forces Q_{C1}, Q_{C2}, Q_{C3}.

(b) Boom-panel structures (see Figure 7.6)

The other way of reacting shear panel edge loads is to attach a flange, or 'boom' to the edge. This might be used, for example, to represent a cantilevered structure, such as an inner fender assembly, with edge members top and bottom to react the shear panel edge loads (see Figure 7.6). The transverse shear stiffness of the panel is

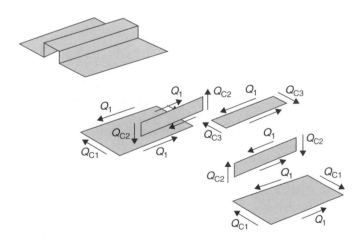

Figure 7.5 Floor in shear.

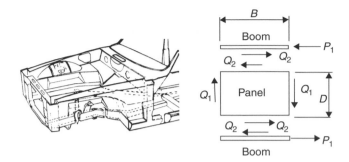

Figure 7.6 Boom-panel assembly.

so much greater than that of the edge members that the contribution of the latter to carrying the transverse load is ignored. Only the axial forces in the edge members or booms are taken into account. The boom is thus a one-dimensional equivalent of the two-dimensional simple structural surface.

The reaction to the complementary shear force Q_2 is balanced (in the structure in Figure 7.6) by an equal and opposite shear force on the edge of the boom. This, in turn, is balanced by an axial force P_1 in the boom. This builds up approximately linearly in response to the shear flow on the edge of the attached panel.

Inspection of the boom/panel assembly reveals that it is acting, overall, as a *beam* (a member which can carry bending and shear). The offset D (the depth of the assembly) between the forces P_1 causes a moment $P_1 D$ at the right-hand edge. An assumption in the simple structural surfaces method (and elsewhere) is that, in a beam composed of booms and panels:

(a) the bending moment in the assembly is reacted by the axial forces in the booms, acting at an offset D from each other;
(b) the shear panel carries all the shear force Q_1.

It is also assumed that the axial force P_1 in flat, thin walled flanges attached to panels is concentrated near to the junction of the panel and flange, due to shear lag effects. (*Note*: If there is no reacting structure on the edge of a shear panel (i.e. a free edge), then the shear forces (and hence shear flow) on that edge must be zero. In the absence of other external moments of couples, the complementary shear forces on the panel would also disappear.)

If a boom is present, then the shear force Q_2 is transferred to the boom via a shear flow, q_2 along the edge (see Figure 7.7). A common simplifying assumption is that q_2 is constant with distance X along the edge, so that $q_2 = Q_2/B$. This is only true for an 'infinitely stiff' boom. In this case, the axial force P_x would build up linearly with x along the boom thus:

$$P_x = q_2 x \qquad (7.6a)$$

Given that $q_2 = Q_2/B$ and $Q_2 = Q_1 B/D$ the bending moment M_X on the assembly at distance X from A is $M_X = P_x D = (q_2 X) D = Q_1 X$ as expected.

In reality the boom is not infinitely stiff axially, and this leads to an effect called 'shear lag' whereby the shear flow q_2 is not constant along the panel edge as depicted in Figure 7.8. The axial force P_x at distance X along the boom now becomes:

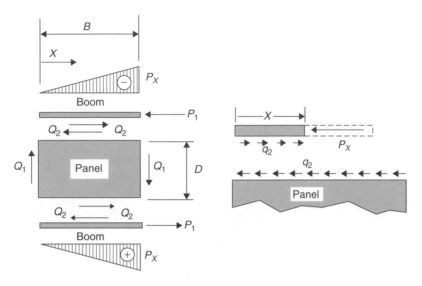

Figure 7.7 Shear flow q_2 and boom force P_X.

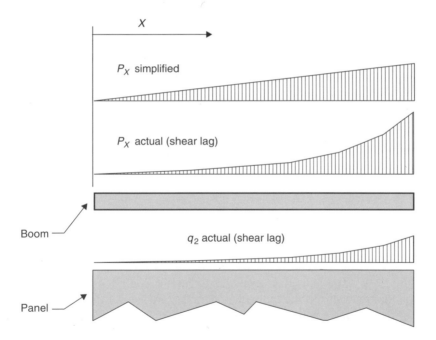

Figure 7.8 Boom force with and without shear lag.

$$P_x = \int_0^B q_2 \, dX \qquad (7.6b)$$

P_x will build up with X as shown, so that the maximum axial force in the boom will be greater than that predicted by the constant shear flow assumption. Further discussion of shear lag is outside the scope of this book, the reader is referred to Megson (1990).

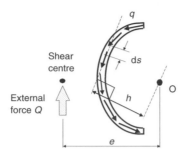

Figure 7.9 Shear centre.

Although the lateral stiffness of the boom is usually ignored in the panel shear direction, the boom may nevertheless be a 'beam' (i.e. with local bending stiffness), and its stiffness may be needed to resist out-of-plane forces, or twist on the assembly (see, for example, the 'sloping inner fender' discussion in Chapter 6, section 6.2.7).

(c) Curved panels

Curved panels are well capable of carrying shear, but if such a panel is to be loaded transversely as part of a boom panel construction, then the load must be applied through the 'shear centre' of the assembly to avoid twisting. The shear centre is the point relative to the cross-sectional shape through which the external load must be applied to balance the torque developed by the shear flows acting around the perimeter of the section (see Figure 7.9).

Taking moments about arbitrary point O in Figure 7.9:

$$Qe = \int hq \, ds$$

where the symbols are as defined in the figure. Readers unfamiliar with this concept are referred to Megson (1990).

7.2.2 Triangulated truss

This is composed of 'pin-ended' members connected in a triangulated arrangement so that they 'lock'. The individual members experience only tension or compression. Even if the joints are stiff, if the individual members are slender then the local bending stiffness of the members will be low, so that their behaviour will approximate to that of pin-ended rods.

A triangulated bay can act as a simple structural surface, in that it can carry a complementary set of edge shear forces Q_1 and Q_2 as shown in Figure 7.10. As with the panel, $Q_1 B = Q_2 D$ to satisfy the requirements of moment equilibrium. If the edges of the triangulated bay are connected to shear panels, forces Q_1 and Q_2 will be fed in as shear flows, $q_1 = Q_1/D$ and $q_2 = Q_2/B$. The axial forces P_1 and P_2 in the edge members will thus vary along their length.

The shear forces Q_1 and Q_2 are reacted by the force P_3 in the diagonal member. This may be seen in the vertically sectioned bay in Figure 7.11(a) by resolving in the

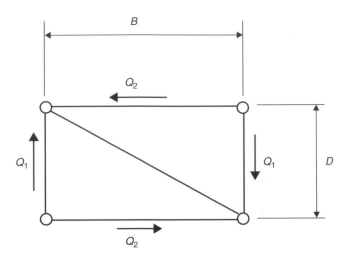

Figure 7.10 Triangulated bay as a simple structural surface.

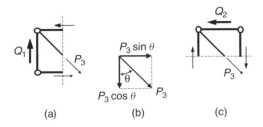

(a)   (b)   (c)

Figure 7.11 Force in diagonal member.

direction of Q_1. If the frame is rectangular, the forces on the horizontal edges have no component in this direction. The only member force which can be resolved into the direction of Q_1 is P_3, in the diagonal Figure 7.11(b). For vertical force equilibrium:

$$Q_1 = P_3 \cos \theta \tag{7.7a}$$

where θ is the angle of the diagonal member as shown in the figure.

Similarly, the horizontal edge force Q_2 is resisted only by the force in the diagonal member. Resolving horizontally on the sectioned surface in Figure 7.11(c):

$$Q_2 = P_3 \sin \theta \tag{7.7b}$$

(b) Triangulated bay structures equivalent to boom-panel assemblies

A multi-bay triangulated truss, acting as a cantilever beam assembly, is shown in Figure 7.12. As before, for vertical equilibrium, the shear force Q at any sectioned surface distance X from the applied load is reacted by the vertical component $P_3 \cos(\theta)$ of the force P_3 in the diagonal member.

Most of the bending moment $M = Qx$ is furnished by the forces P_2 and P_4 in the *horizontal* ('flange') members, acting at offset D from each other. For the second and

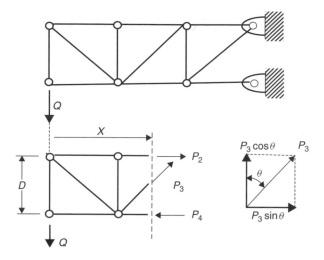

Figure 7.12 Multi-bay triangulated truss.

subsequent bays, $M \approx P_2 D$. The difference between P_2 and P_4 is accounted for by the horizontal component $P_3 \sin(\theta)$ of the force in the diagonal member. Such a structure (single or multiple bay) can be seen as equivalent to a boom-panel assembly. The top and bottom members act as the 'booms' and the diagonal is analogous to the shear panel.

For a laterally loaded single bay rectangular triangulated truss (or the first bay in a multiple bay truss) as shown in Figure 7.13, the forces in members 1 to 5 are:

$$P_1 = Q \qquad \qquad \text{(Tension)}$$

$$P_2 = P_3 \sin \theta = Q \tan \theta \qquad \text{(Tension)}$$

$$P_3 = Q / \cos \theta \qquad \qquad \text{(Compression)}$$

$$P_4 = 0$$

$$P_5 = P_3 \cos \theta = Q \qquad \text{(Tension)} \qquad\qquad (7.8)$$

Again, the shear force is carried by the diagonal.

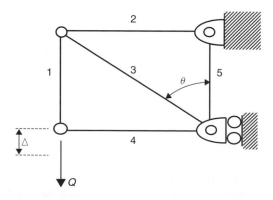

Figure 7.13 Single bay triangulated truss.

Stiffness of triangulated bay

Strain energy methods may be used to calculate deflections. These methods are explained in Gere and Timoshenko (1991). For axially loaded members only, the total strain energy of the frame (treating the member loads as constant) is:

$$U = \sum \frac{P^2 L}{2AE}$$

Where A and L are the length and cross-sectional area of each member, and E is Young's modulus for the material. In this case, for the *total* bay:

$$U = \underbrace{\frac{P_1^2 L_1}{2A_1 E}}_{\text{Member 1}} + \underbrace{\frac{P_2^2 L_2}{2A_2 E}}_{\text{Member 2}} + \underbrace{\frac{P_3^2 L_3}{2A_3 E}}_{\text{Member 3}} + \underbrace{\frac{P_4^2 L_4}{2A_4 E}}_{\text{Member 4}} + \underbrace{\frac{P_5^2 L_5}{2A_5 E}}_{\text{Member 5}} \qquad (7.9a)$$

But this is the strain energy of the *whole frame*. Arguing that the *shear* force Q_1 is carried by the diagonal (member 3) *alone*:

$$U_{\text{SHEAR}} = \frac{P_3^2 L_3}{2A_3 E} = \frac{Q_1^2}{\cos^2 \theta} \frac{L_3}{2A_3 E} \qquad (7.9b)$$

Using Castigliano's theorem, the deflection Δ as shown in Figure 7.13 is:

$$\Delta = \frac{\partial U}{\partial Q_1} = \frac{2Q_1 L_3}{2A_3 E \cos^2 \theta} \qquad (7.10)$$

The remainder of the terms in the energy equation refer to the 'edge members' (1, 2, 4, 5) acting as 'booms'. These provide overall bending stiffness of the bay.

Thus, the equivalent 'shear panel' stiffness K of the triangulated single bay, using the force in the diagonal only, is:

$$K_{\text{EQUIV SHEAR}} = \frac{Q_1}{\Delta_1} = \frac{A_3 E \cos^2 \theta}{L_3} \qquad (7.11)$$

The volume (related to the weight) of the bay is:

$$VOL = A_3 L_3 \quad \text{(diagonal member only)} \qquad (7.12a)$$

or

$$VOL = A_1 L_1 + A_2 L_2 + A_3 L_3 + A_4 L_4 + A_5 L_5 \quad \text{(whole bay)} \qquad (7.12b)$$

7.2.3   Single or multiple open bay ring frames

(a) Single open frame as simple structural surface

A rectangular open ring frame with stiff edge members and stiff corner joints can behave as a simple structural surface. This arrangement is much more flexible in overall shear than a continuous panel because it derives its global stiffness from local bending in the edge members. Large local bending moments, about axes *normal* to the plane of the frame, are present in the edge beams. Since the edge members are

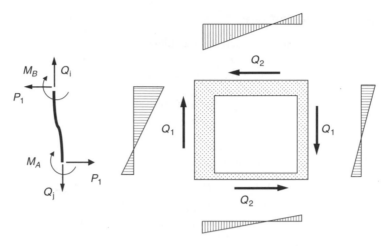

Figure 7.14 Single bay ring frame.

in contraflexure, the moments are greatest at the corners, so that particular design attention should be paid to the bending stiffness of the corner joints (see Figure 7.14).

The distribution and magnitude of the bending moments depends on the relative stiffness in bending of the edge members.

Symmetric ring frame as simple structural surface

If the frame has two planes of symmetry, then the corner moments are all the same, and the zero bending moment points are in the middle of the sides, and the shear forces on opposite sides of the frame are equal as shown in Figure 7.15 This leads to a simplification in calculating the internal loads and the deflections of the frame.

If the symmetric frame is sectioned at its two planes of symmetry, half-way across its height and width, then, from symmetry considerations:

(a) the local bending moments in the edge members at the planes of symmetry will
be zero;

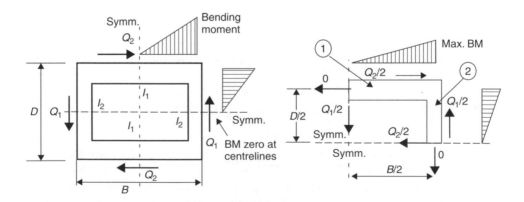

Figure 7.15 Forces on symmetrical frame under pure shear case.

(b) the local shear forces in the horizontal (as seen in Figure 7.15) edge members will be $Q_1/2$. This is because the force Q_1 will be shared equally by these members as it passes across the frame;

(c) the shear forces in the vertical edge members will be $Q_2/2$, for similar reasons.

In the pure shear case, if the forces Q_1 and Q_2 are the result of constant shear flows q_1 and q_2, then the forces applied by these shear flows to the quarter structure (i.e. divided at the planes of symmetry) will be $Q_1/2$ and $Q_2/2$ on the vertical and horizontal limbs respectively as shown in Figure 7.15. Applying horizontal, vertical and moment equilibrium conditions to the quarter frame will reveal the loads as shown in Figure 7.15(b). From this it can be seen that the maximum bending moment in the edge members at any of the corners of the frame is:

$$M_{\text{MAX.}} = Q_1 B/4 = Q_2 D/4 \tag{7.13}$$

(b) Open bay ring frame equivalents to boom-panel assembly

A multiple bay open ring frame (known as a verandeel truss) can be seen in Figure 7.16. Such an arrangement is capable of carrying lateral loads as shown.

Considering the truss sectioned at some distance X from the load there will be a global bending moment $M = QX$, as well as a shear force Q acting at the section. By analogy to the shear panel/boom structure:

(a) The global bending moment is carried mainly by the couple $P_1 D$ due to offset axial forces P_1 in the top and bottom members. The local bending moments M_1 and M_2 in the individual edge beams are small compared with this couple.

(b) There is no shear member, so that the shear force Q must be carried as lateral forces in the top and bottom members, which now act as independent beams. This leads to local bending moments in these members. The frame is thus comparatively flexible overall in shear.

Shear stiffness of single bay symmetric open bay truss structure

A single bay open ring frame truss, simply supported and loaded laterally with force Q_1, is shown in Figure 7.17. This is analogous to a boom-panel assembly, and so carries

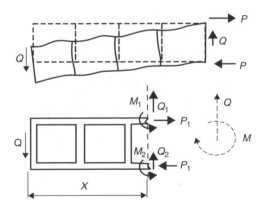

Figure 7.16 Multiple bay open frame truss.

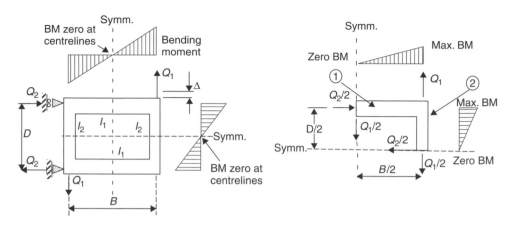

Figure 7.17 Single bay symmetric open frame truss.

an overall bending moment as well as the overall shear force Q_1. At the restrained end, the overall bending moment on the truss is Q_2D.

Symmetry of the bending moment distribution within the edge members and of member shear forces is the same as for the pure shear case (Figure 7.15). The slight differences in the support conditions cause the axial forces in the members to be differently distributed, as shown in Figure 7.17.

Using the notation in Figure 7.17 the bending moment M_1 in member 1 distance X from the plane of symmetry is:

$$M_1 = Q_1X/2 \tag{7.14a}$$

and in member 2, distance Y from the plane of symmetry, bending moment M_2 is:

$$M_2 = Q_2Y/2 = Q_1BY/(2D) \quad \text{(since } Q_2 = Q_1B/D) \tag{7.14b}$$

The shear deflection of the truss may be determined using a strain energy method. Readers unfamiliar with this approach are referred to Gere and Timoshenko (1991).

The strain energy $U_{1/4}$ of the quarter frame is:

$$U_{1/4} = \int_0^{B/2} M_1^2/(2EI_1)\,dX + \int_0^{D/2} M_2^2/(2EI_2)\,dY$$

where E = Young's modulus of the material and I_1 and I_2 are the second moments of area of members 1 and 2.

Since there are four quarter frames in the whole assembly, the total strain energy U is:

$$U = 4U_{1/4} = Q_1^2B^2\{B/I_1 + D/I_2\}/(48E) \tag{7.15}$$

From Castigliano's theorem, the shear deflection Δ is then given by:

$$\Delta = \partial U/\partial Q_1 = Q_1B^2\{B/I_1 + D/I_2\}/(24E) \tag{7.16}$$

The shear stiffness K of the open bay truss is:

$$K = (24E)/\{B^2(B/I_1 + D/I_2)\} \quad \text{(where } Q_1 = K\Delta) \tag{7.17}$$

Symmetric frame: summary

	Internal loads in a symmetrical ring frame in shear	
	Member 1	Member 2
Axial force	$Q_2/2 = Q_1 B/2D$	$Q_1/2 = Q_2 D/2B$
Shear force	$Q_1/2 = Q_2 D/2B$	$Q_2/2 = Q_1 B/2D$
Max. bending moment	$M_0 = Q_1 B/4$ (All corners)	
Deflection Δ in the direction of Q_1	$\dfrac{Q_1 B^2}{24E}\left(\dfrac{B}{I_1} + \dfrac{D}{I_2}\right)$	
Stiffness K in the direction of Q_1	$\dfrac{24E}{B^2\left(\dfrac{B}{I_1} + \dfrac{D}{I_2}\right)}$	

7.2.4 Comparison of stiffness/weight of different simple structural surfaces

The weights of three structural surfaces all of the same overall size and shear stiffness are calculated and compared. These are (see Figure 7.18):

(a) simple flat shear panel;
(b) pin-jointed triangulated frame;
(c) stiff-jointed open ring frame.

General data:
Material: Steel   Young's modulus   210 000 N/mm²
                   Shear modulus   80 000 N/mm²
Size: Width   500 mm
                   Depth   500 mm

Because of the relatively low stiffness of ring frames, the baseline for *stiffness* is chosen to be the ring frame. The cross-sectional dimensions of the edge members are chosen to be typical of a car body beam section as follows:

Baseline ring frame:
Edge beam: Cross-section $100 \times 100 \times 2.0$ mm square hollow section
            Second moment of area I 1.255×10^6 mm⁴
            Cross-sectional area A 800 mm²

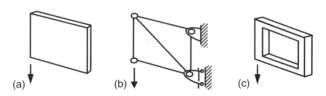

(a)   (b)   (c)

Figure 7.18 Three structural surfaces compared.

Table 7.1 Alternative structures with the same shear stiffness of 25 300 N/mm

Shear stiffness K (N/mm)	Structural surface type	Dimensions (mm) and cross-section properties	Approx. total volume (mm³)	Comparative volume (hence: weight)
25 300	Shear panel	$500 \times 500 \times 0.316$	79 000	1
25 300	Triangulated frame	Square section $A = 170\,\text{mm}^2$	460 190	5.83
		Cross-section size: $35 \times 35 \times 1.2\,\text{mm}$	(diagonal only: 120 190)	(1.52)
25 300	stiff-jointed frame	Square section $I = 1.255 \times 10^6\,\text{mm}^4$ $A = 800\,\text{mm}^2$ Cross-section size: $100 \times 100 \times 2\,\text{mm}$	1 040 000	13.16

Shear stiffness of ring frame:

$$K = (24E)/\{B^2(B/I_1 + D/I_2)\} = 25\,300\,\text{N/mm}$$

Perfect joints and constant section around the perimeter of the frame are assumed.

The other structures are sized to give the same shear stiffness. The cross-sectional sizes are given in Table 7.1. This gives a very thin shear panel, which although of acceptable stiffness, would probably require added stiffeners to prevent shear buckling.

The baseline for *weight* is the shear panel, because of its high weight efficiency. In the comparison table, the volumes (and hence the weights) of the other structures are divided by the volume of the shear panel.

All of the members of the triangulated section are assumed to be of a single size.

The results of the comparison are shown in Table 7.1.

These results are for one size and section and one set of dimensions of each surface type. Other sizes can be compared on a stiffness to weight basis using the stiffness formulae developed in sections 7.2.1, 7.2.2 and 7.2.3.

In any such simplified calculation, careful attention should be paid to the assumptions used. For example, the above calculations on the ring frames assume corner joints which are perfectly stiff in bending. This is rarely the case in reality, and corner joint flexibility can dominate the overall shear stiffness of the ring beam, particularly if the joints are poor.

7.2.5   Simple structural surfaces with additional external loads

In the SSS vehicle body load-path calculations, some panels were seen to experience extra external forces as well as edge loads, and this affected the values of the edge loads.

An example is the front bulkhead of the standard sedan in the torsion load case (see Figure 7.19, and Figure 5.12, Chapter 5). This is subject to inner fender shear reactions R_F, causing couple $R_F S_F$ in addition to edge forces Q_1 and Q_2. In following *Global* vehicle load paths, it was sufficient to consider only the *net* edge force Q_1 on the top

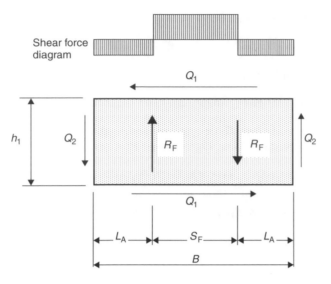

Figure 7.19 Bulkhead panel.

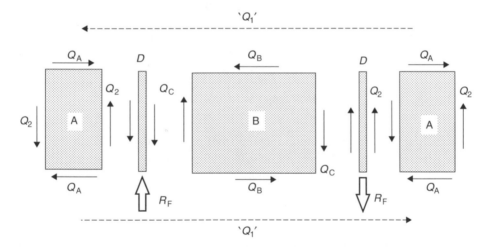

Figure 7.20 Panel breakdown for bulkhead.

and bottom of this panel. However, forces R_F will cause varying shear force across the panel as shown in Figure 7.19.

In order to analyse this (or similar panels) for detailed shear forces or shear flows, a more detailed panel breakdown is needed into outer panels A and central panel B as shown in the Figure 7.20. Forces R_F are passed from the inner fender webs to the bulkhead at the small strips D of the bulkhead. These strips will be treated as booms to allow force balance calculations between panels A and B and the inner fender webs.

Panels A and B may now be seen to be in pure complementary shear. Using the notation in Figures 7.19 and 7.20, and taking a moment balance on one of the panels A, the complementary shear forces on Q_A are given by:

$$Q_2 L_A = Q_A h_1 \quad \text{so that} \quad Q_A = Q_2 L_A / h_1$$

For vertical force equilibrium in the booms D:

$$Q_2 + Q_C - R_F = 0 \text{ giving inner panel (B) shear force} = Q_C = -Q_2 + R_F$$

(i.e. there is a change of shear force direction and magnitude at D).

The complementary shear force Q_B on panel B is given by:

$$Q_B h_1 = Q_C S_F \quad \text{thus} \quad Q_B = Q_C S_F / h_1 = (R_F - Q_2) S_F / h_1$$

Shear flow on panels A and B

The shear flow q_A on panel A is:

$$q_A = Q_A / L_A$$

Substituting for Q_A from above:

$$q_A = \{Q_2 L_A / h_1\} / L_A = Q_2 / h_1 = q_2$$

as expected from the complementary shear flow rule for shear flows q_A and q_2.

Similarly, on panel B, shear flow q_C is equal to complementary shear flow q_B as follows:

$$q_c = \frac{Q_C}{h_1} = \frac{\left(\dfrac{Q_B h_1}{S_F} \right)}{h_1} = \frac{Q_B}{S_F} = q_B$$

The shear flow reverses direction on moving from panel A to panel B, so that forces Q_A and Q_B are different in direction and magnitude. This is as expected from the shear force diagram of the panel. Summing the forces along the top of the panel:

$$Q_1 = Q_B - 2Q_A$$

In other words, Q_1 is the *net* force acting along the top of the bulkhead. Nevertheless, the original moment equilibrium equation for the bulkhead as a whole still applies, viz.:

$$Q_1 h_1 + Q_2 B = R_F S_F$$

where B is the total width of the bulkhead.

7.3 In-plane forces in sideframes

Vehicle sideframes are in a particular class of structural surface involving ring frames (Figure 7.21). The edges of these are often supported by multiple pillars, which, with the header and rocker, form multiple ring frames. The total edge force on the top edge of the sideframe Q_{TOTAL} acts on the cantrail, and is reacted by the sum of the local shear loads in the individual pillars (Figure 7.21). The pillars may or may not apply

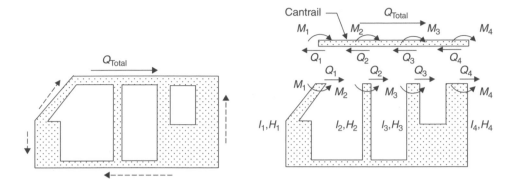

Figure 7.21 Pillar/cantrail loads in sideframe.

local bending moments to the cantrail at their attachment points, depending on the quality of the structural joints (see later).

Such frames are statically indeterminate (three redundancies per two-dimensional ring), so that exact determination of shear forces and hence bending moments in each individual pillar is complicated. A finite element model could be used, or, at the early conceptual design stage, simplified rough estimates can be made as follows.

7.3.1 Approximate estimates of pillar loads in sideframes

(a) Pillar with rigid end joints

A rough estimate of pillar forces, for a pillar with good joints at its ends, may be made by assuming that they are built in to perfectly stiff structures top and bottom. For uniform pillars under shear in such a situation, the bending moment diagram is as shown in Figure 7.22 with zero bending moment at the middle. The pillar thus behaves as two cantilever beams 'tip to tip'.

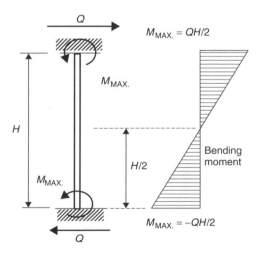

Figure 7.22 Bending moment in pillar with stiff joints.

Thus the maximum bending moment at the ends of the pillar is:

$$M_{\text{MAX.}} = Q\frac{H}{2} \tag{7.18}$$

where the notation is defined in Figure 7.22.

Also (using the cantilever bending formula for the two halves of the pillar):

$$\text{deflection } \Delta = 2\left(\frac{Q\left(\frac{H}{2}\right)^3}{3EI}\right) = \frac{QH^3}{12EI} \tag{7.19}$$

where E = Young's modulus, I = second moment of area of pillar section and the other symbols are as defined in Figure 7.22.

For multiple pillars attached to a stiff header (Figure 7.21), the compatibility condition, based on the assumption of rigid, inextensible structures (i.e. waistrail and cantrail) attached to the ends of the pillar, is that the lateral deflections are the same for all of the pillars:

$$\Delta = \Delta_{\text{PILLAR 1}} = \Delta_{\text{PILLAR 2}} = \Delta_{\text{PILLAR 3}} \cdots \tag{7.20}$$

Also for equilibrium of the cantrail the total edge force Q_{TOTAL} on the surface is:

$$Q_{\text{TOTAL}} = Q_1 + Q_2 + Q_3 + \ldots = \sum Q_i \tag{7.21}$$

where Q_1, Q_2, Q_3, Q_i etc. are the lateral forces on pillars 1, 2, 3, i, etc.

Rewriting (7.19) for the ith pillar as:

$$Q_i = 12EI_i\Delta/H_i^3 \tag{7.22}$$

The deflection Δ is the same for all pillars, so combining equation (7.21) with (7.22) gives the following for Q_{TOTAL}:

$$Q_{\text{TOTAL}} = (12E\Delta)\{I_1/H_1^3 + I_2/H_2^3 + I_3/H_3^3 \ldots\} = (12E\Delta)\sum\{I_i/H_i^3\} \tag{7.23}$$

Combining equations (7.22) and (7.23) and eliminating common terms gives an expression for the shear force Q_j in the jth pillar:

$$Q_j = Q_{\text{TOTAL}}\left(I_j/H_j^3\right) \Big/ \left(\sum\{I_i/H_i^3\}\right) \tag{7.24}$$

For example, the shear force Q_2 in pillar 2 of a three pillar sideframe subject to total cantrail force Q_{TOTAL} is:

$$Q_2 = Q_{\text{TOTAL}}(I_2/H_2^3)/\{I_1/H_1^3 + I_2/H_2^3 + I_3/H_3^3\}$$

(b) Poor joint at one end of pillar

If the joint at one end of a pillar is poor in bending, then it can be approximated by a pin joint (no bending stiffness at this site – see Figure 7.23).

In this case, the maximum bending moment $M_{\text{MAX.}}$, and the deflection Δ are:

$$M_{\text{MAX.}} = QH \quad \Delta = \frac{QH^3}{3EI} \tag{7.25}$$

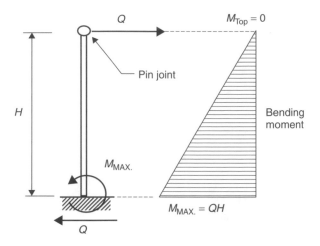

Figure 7.23 Bending moment in pillar with a pin joint at the top.

These expressions should be used for the pillar in the above formulae. Note that this is now a single, longer cantilever so that $M_{\text{MAX.}}$ is twice as large as for the stiff-jointed pillar. The pin-jointed pillar has only a quarter of the lateral stiffness of the stiff-jointed one.

(c) Mixture of joint conditions at the ends of the pillars
Often there are cases with a mixture of joint conditions on the ends of the pillars. For example, consider a sideframe with stiff joints at both ends of the B- and C-pillars, but with a poor joint at the top of the A-pillar.

For the A-pillar:
$$Q_A = 3EI_A/H_A^3$$

For the B- and C-pillars:
$$Q_B = 12EI_B/H_B^3 \text{ and}$$
$$Q_C = 12EI_C/H_C^3$$

The expression for the shear force in the A-pillar is:

$$Q_A = Q_{\text{TOTAL}}(3I_A/H_A^3)/\{3I_A/H_A^3 + 12I_{Br}/H_B^3 + 12I_C/H_C^3\}$$

where the subscripts A, B, C refer to the A-, B- and C-pillars, respectively.
   For the B-pillar, the shear force is:

$$Q_B = Q_{\text{TOTAL}}(12I_B/H_B^3)/\{3I_A/H_A^3 + 12I_{Br}/H_B^3 + 12I_C/H_C^3\}$$

(d) Slanted pillars
If a pillar is slanted, Pawlowski (1969) suggested that, for a first estimate of pillar forces, the total true length be used in the above formulae. Once the load share Q on the pillar is known from this, it can be resolved into axial and lateral force components in the pillar. The axial pillar force P_{AXIAL} and the bending moment M in the pillar can then be calculated accordingly.

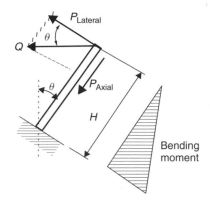

Figure 7.24 Force components on slanted pillar.

Using the notation in Figure 7.24:

$$P_{\text{AXIAL}} = Q \sin(\theta)$$

$$P_{\text{LATERAL}} = Q \cos(\theta)$$

$$M_{\text{MAX.}} = HQ \cos(\theta) \qquad \text{(for pillar with pin joint at top)}$$

$$M_{\text{MAX.}} = H(Q/2) \cos(\theta) \quad \text{(for pillar with stiff joint at top)}$$

(7.26)

(e) Pillars shared by adjacent structural surfaces

Certain pillars (e.g. the A-pillar) share shear forces from adjacent mutually perpendicular ring beams. In this case, the total stress is the sum of those caused by loads in each frame. In the example shown in Figure 7.25, it is assumed that the joint at the top of the A-pillar is flexible in bending. Using the notation in the diagram and taking H as the true length of the pillar:

$$M_{XX} = P_{\text{LATERAL }1}H \quad (P_{\text{LATERAL }1} = \text{component of lateral pillar force} \\ \text{from sideframe})$$

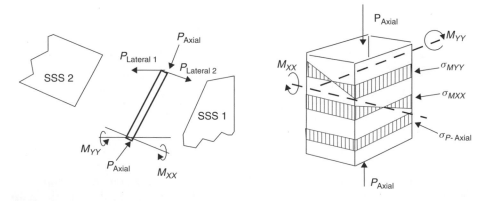

Figure 7.25 A-pillar shared between sideframe and windshield.

$$M_{YY} = P_{\text{LATERAL } 2} H \quad (P_{\text{LATERAL } 2} = \text{component of lateral pillar force}$$
$$\text{from windshield)}$$

$P_{\text{AXIAL}} = $ sum of axial components of force from sideframe and windshield

The stresses caused by M_{XX} and M_{YY} may be calculated from engineers' bending theory, and added to that resulting from P_{AXIAL}. Since M_{XX} and M_{YY} are about mutually perpendicular axes, the stresses will add in some corners and subtract in others. The maximum stress will be in the corner where all three stresses have the same sign.

7.4   Loads normal to surfaces: floor structures

Floors are subject to loads normal to their plane. Under such circumstances they do not act as simple structural surfaces. A flat sheet of thin material such as the floor panel is very flexible for out-of-plane loads, so that in the SSS method its out-of-plane stiffness is ignored. The floor is stiffened against out-of-plane loads by added beam members arranged into a planar framework called a grillage. A typical car floor structure is shown in Figure 7.26.

7.4.1   Grillages

A true grillage is a flat frame loaded normal to its plane. In such a structure the only active forces in each individual member are:

Figure 7.26 Typical car floor structure.

Normal force (normal to plane of grillage)

Bending moment (about in-plane axis)

Torque (about in-plane axis)

The major grillage members in the passenger compartment floor consist principally of the 'transmission tunnel', one or more cross-members, the rockers (as part of the sideframe) and the bulkheads at the ends of the compartment. These members, in idealized form, are shown in Figures 7.27, 7.28 and 7.29. Even in modern front wheel drive vehicles, a longitudinal 'tunnel' member will almost always still be found, and this underlines its structural significance. A better name for the member in this case would be the 'services tunnel' since it often carries 'services' such as brake pipes, handbrake cable, etc.

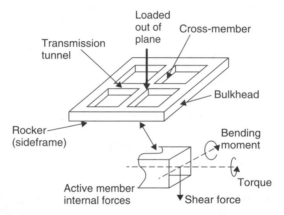

Figure 7.27 Grillage structure.

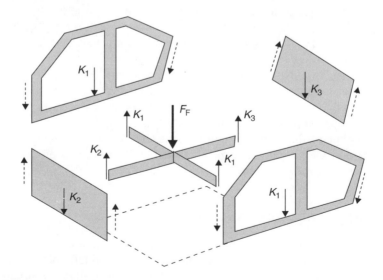

Figure 7.28 Load distribution in floor grillage.

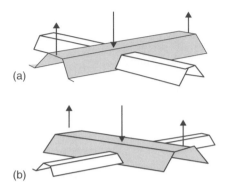

Figure 7.29 Alternative floor arrangements. (a) Stiff tunnel and flexible crossmember (load transmitted to ends); (b) flexible tunnel and stiff crossmember (load transmitted to sides).

The grillage members in the car floor can consist of two basic forms:

(a) Members integral to the floor panel are created by forming part of the floor panel itself into beam-like shapes. Examples of this are the (longitudinal) transmission/services tunnel and the (transverse) step in the floor at the rear seat.
(b) Added members, consisting of separate beams of top-hat section formed out of sheet metal, are welded onto the floor panel. The cross-members are usually of this form.
(c) Bulkheads at the ends of the passenger compartment may be thought of as part of the structure carrying the out-of-plane forces from the floor. Being very deep sheet metal panels, these work as shear panels rather than as beams.

7.4.2 The floor as a load gatherer

The aim of the floor is to carry the local applied loads from their point of application to the major structural components of the vehicle, such as the sideframes. Point loads, such as passenger seat reactions, are usually fed straight into local, probably minor, beam members. Distributed payloads frequently rest on the floor panel itself, which will need local reinforcement (in the form of 'swages' or corrugations – see later). The local load is carried along the swages to adjacent beam members (for example, cross-members). The local members will transfer it to larger members (such as the transmission/services tunnel) and so forth until it reaches the sideframe. The progression of the loads through the structure follows the pattern shown in Table 7.2.

The precise route of the loads through the structure depends on the orientation and relative stiffness of the various floor members and on the construction details such as joint quality.

7.4.3 Load distribution in floor members

The share of forces in the different floor members can have a significant effect on the shear force and bending moment distribution in the vehicle sideframe.

Consider, for example, a floor with a single cross-member attached at each side to the rocker members, and a transmission/services tunnel attached at each of its ends to

Table 7.2

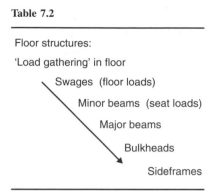

Floor structures:

'Load gathering' in floor

Swages (floor loads)

Minor beams (seat loads)

Major beams

Bulkheads

Sideframes

bulkheads at the ends of the passenger compartment (Figure 7.28). The two members cross at a centre joint. As a simplification, let us assume that all of payload force F is applied at the central joint. The reactions K_1 at the ends of the cross-member will be the same because of symmetry. The cross-member probably will not cross the tunnel at its centre, so that the reactions K_2 and K_3 at its ends will not be the same.

(a) Perfect joints at centre intersection

In the case of perfect joints between the members, the distribution of forces will be in proportion to the stiffness of these members. If the transmission/services tunnel is stiff (which it usually is, being of deep formed section) and the cross-member is flexible, then the tunnel will carry nearly all of the load along its length to the compartment bulkheads as shown in Figure 7.29(a). In the converse case, with a very stiff cross-member and a flexible tunnel member, most of the load will be conducted by the cross-member straight to the sideframes (Figure 7.29(b)).

If the tunnel and the cross-member both have stiffness of the same order, then an approximate approach to the calculation of load share in the floor members is to treat these members as cantilevers joined at the centre – see Figure 7.30. This neglects the slope of the members at the centre joint but is sufficiently accurate to demonstrate floor load share, particularly if the centre joint is near the centre of both members.

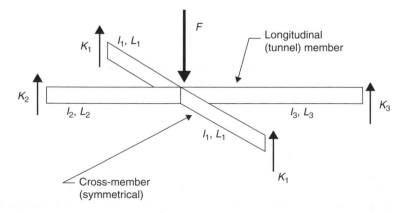

Figure 7.30 Load distribution in the floor.

The outer ends of the members are assumed to be 'pin-jointed' so that they will not develop moments. This is reasonable given the likely nature of the joints at the ends. This will be discussed later in this chapter. An assumption in simple structural surfaces method calculations is to neglect the torsion loads in the floor members, in which case all the load is carried by local bending. This is reasonable for integral floor members, which are of open section and hence of low torsion stiffness.

If the centre deflects Δ while the ends remain stationary then:

$$\Delta = K_1 \frac{L_1^3}{3EI_i} = K_2 \frac{L_2^3}{3EI_2} = K_3 \frac{L_3^3}{3EI_3}$$

where K_1 = the load carried by member 1, L_1 = length of member 1, E = Young's modulus and I_1 = Second moment of area of member i.

But, from equilibrium considerations the total load is the sum of the reactions. The two sides of the cross-member (member 1) are identical from symmetry. Thus:

$$F = 2K_1 + K_2 + K_3 = \left\{ \frac{2I_1}{L_1^3} + \frac{I_3}{L_2^3} + \frac{I_3}{L_3^3} \right\} 3E\Delta$$

Taking a ratio of K_i to F gives, in the general case:

$$K_i = \frac{\left(\dfrac{I_i}{L_i^3} \right)}{\left(\dfrac{2I_1}{L_1^3} + \dfrac{I_2}{L_2^3} + \dfrac{I_3}{L_3^3} \right)} F \tag{7.27}$$

i.e.
$$K_1 : K_2 : K_3 = \frac{I_1}{L_1^3} : \frac{I_2}{L_2^3} : \frac{I_3}{L_3^3}$$

If the joints between the members are perfect, then the loads will be distributed (approximately):

- in proportion to the member I values;
- inversely proportional to the cube of the member lengths.

Special case (a)   If the lengths of all the limbs are the same (i.e. the joint is truly at the centre of the floor), then $L_1 = L_2 = L_3$ and:

$$K_i = \frac{I_i}{(2I_1 + I_2 + I_3)} F \tag{7.28}$$

Thus the load is distributed in proportion to the second moments of area of the limbs in this case.

Special case (b)   If the second moments of area of the sections of all the limbs are the same (i.e. the tunnel and the cross-member have (approximately) the same section) then $I_1 = I_2 = I_3$ and:

$$K_i = \frac{\left(\dfrac{1}{L_i^3}\right)}{\left(\dfrac{2}{L_1^3} + \dfrac{1}{L_2^3} + \dfrac{1}{L_3^3}\right)} F \qquad (7.29)$$

so that in this case the load is distributed in proportion to the inverse of the cube of the lengths of the limbs. Thus load share is very sensitive to length and long members will only carry a very small proportion of load L.

(b) Effect of joint flexibility on load distribution

The joints between the members can have a significant effect on load distribution.

If a mid-floor joint interrupts the bending stiffness of the cross-member more than that of the longitudinal, then most of the load will be transmitted by the longitudinal transmission/services tunnel and vice versa. This is similar to the effects depicted in Figure 7.29(a) and (b), respectively.

A typical case of load interruption in the cross-member might be where the transmission/services tunnel is continuous and the cross-member is attached to its sides (Figure 7.31). Bending from the cross-member is only resisted by out-of-plane loads in the walls of the tunnel, with consequent low bending stiffness. The overall effect is close to that of the cross-member being pinned to the sides of the tunnel. Bending moments in the cross-member are disrupted, but shear forces are transmitted between the cross-member and the tunnel. If a load F is applied to the central joint, the cross-member will not carry any of it because of this, and all the load will be transferred along the transmission/services tunnel. However, seat loads applied part-way along the cross-member will be transferred to the members at its ends, namely the tunnel and the rocker panels.

There are several possible corrective measures for this situation (see Figure 7.32). They are as follows:

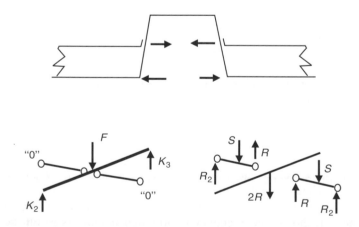

Figure 7.31 Flexible joint between cross-member and transmission/services tunnel.

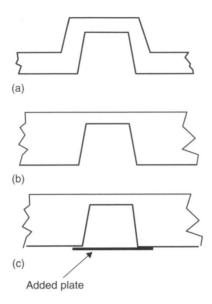

Figure 7.32 Methods of stiffening the cross-member to tunnel joint.

(a) The cross-member can be made continuous by allowing it to run up and over the transmission/services tunnel as in Figure 7.32(a). This usually means that the cross-member is of much smaller section than the tunnel, so that the load share ratio outlined in equation (7.29) will still favour the transmission/services tunnel to carry most of the load.

(b) The cross-member can be made deeper than the transmission/services tunnel, so that a small amount of the cross-member section is carried across the top of the tunnel, as in Figure 7.32(b). Since this carryover is usually shallow, it is fairly flexible in bending, so that the improvement in overall bending stiffness of the cross-member is small.

(c) The open underside of the transmission/services tunnel can be reinforced with a plate attached underneath the floor as in Figure 7.32(c). This might be a transmission steady bracket, for example. For this method to be effective, the top surface of the cross-member must align at least with the top of the tunnel, to ensure a good in-plane load path at this point also (for the local forces caused by the reaction to the cross-member bending moment). The shear stiffness across the tunnel will be quite low, because of 'lozenging'.

The outer end of the cross-member is also often joined onto the unsupported sheet metal on the sides of the of the rocker beam. This gives a joint which closely resembles a pin joint at this site also.

Another site where interruption of stiffness of a cross-member by the transmission/services tunnel occurs is in the step in the floor at the front of the rear seat. The latter provides quite a substantial cross-member. However, it is usually almost completely penetrated by the transmission/services tunnel at the centre, giving low bending stiffness at this point. A common correction for this is to run a continuous added top-hat section member all the way across the top of the seat step. This is often included for crashworthiness purposes, but it also has the effect of reinforcing the seat step in bending.

7.4.4   Swages and corrugations

The out-of-plane stiffness of a thin panel may be increased locally by swages (i.e. impressed grooves). In effect, this entails the local addition of 'beams' to the panel increasing its 'apparent depth'. If the swages merge together as a series of adjacent members, then the panel is referred to as 'corrugated'.

Basic properties of individual swages (see Figure 7.33):

(a) Swages only increase the bending stiffness of the panel about the axis *transverse to the swage*.
(b) Bending stiffness is negligible about the axis *parallel to the swage*.
(c) The out-of-plane panel load will be transferred *along* the swages to their ends by local bending.
(d) The in-plane *membrane* stiffness of the swaged panel is negligible in the direction *normal* to the swage. This is because such loading causes local bending of the sheet about an in-plane axis due to the offset of the panel load from the top of the swage.
(e) For greatest transverse stiffness of the panel, the swages should run across the shortest width of the panel.
(f) Where two swages cross each other, the bending stiffness of both swages is reduced locally. This should be avoided where possible.
(g) A local flat area of sheet metal, between the end of the swage and the supporting floor beam, leads to high local transverse shear flexibility, and can significantly reduce the overall stiffening effect of the swage.

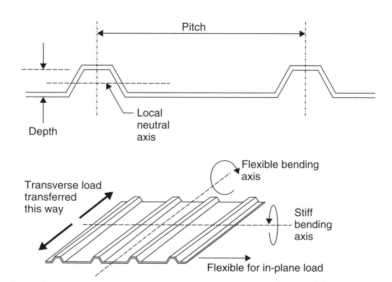

Figure 7.33 Swages.

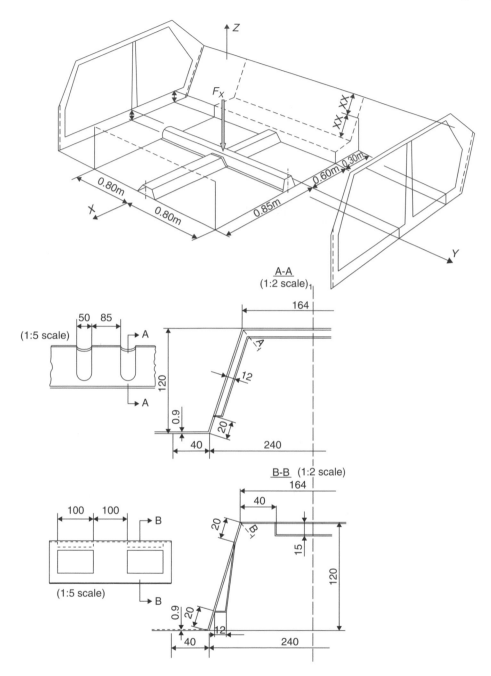

Figure 7.34 Alternative arrangements for longitudinal tunnel in the car floor.

In addition to local stiffening for transverse load transfer, swages and corrugations can be used to increase the vibration frequency of panels, and to reduce the tendency for the panel to buckle. Both of these are due to the increase of local bending stiffness of the panel by the swages.

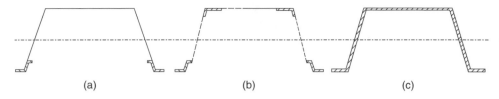

Figure 7.35 Effective cross-sections of transmission/services tunnel.

Interaction of swages and floor-beam members

The swages carry the normal floor load locally to adjacent floor beam members. The beams then carry the accumulated load to the adjacent structural surfaces.

If floor beam members themselves are swaged, this can affect their bending performance. Thus transverse swages can interrupt the longitudinal stresses which carry bending in the beam.

Pawlowski (1986), illustrated this with an example. Three different schemes were compared for local panel stiffening in the central transmission/services tunnel of a typical car structure. The alternative designs for the tunnel member were (see Figure 7.34):

(a) with a series of lateral stiffening swages;
(b) with a series of local rectangular depressions;
(c) smooth.

The lateral swages (a) completely interrupt a large part of the member cross-section, giving the effective section in Figure 7.35(a). The depressions (b) interrupt part of the longitudinal integrity of the cross-section, leaving the effective section in (Figure 7.35(b)). The smooth tunnel has the effective section in Figure 7.35(c).

From equation (7.29), the distribution of the out-of-plane floor load F would be in proportion to the effective I-values of the members. Assuming a cross-member of reasonable dimensions, then for scheme (a) almost none of the floor load would be carried by the transmission/services tunnel. For schemes (b) and (c) the tunnel has retained reasonable bending stiffness, and so it would carry a larger proportion of the floor load.

Application of the SSS method to an existing vehicle structure

Objectives

- To demonstrate the use of the SSS method on a real vehicle body.
- To ensure the structure has satisfactory load-path continuity.
- To determine loads in beams, panels and subassemblies.

8.1 Introduction

While the SSS method is most effectively applied during the conceptual stage, it may also be used to help resolve issues on existing structures. Exercising the method on existing structures can also be used to assess questions such as 'What if we widened or lengthened this vehicle body?', 'What if we changed how the suspension loads are introduced to the body?' For example, it is proposed to modify the existing base platform to be capable of supporting both mid and large size vehicle structures of the same vehicle type. Changes in length, width and vehicle mass will need to be considered. First the base existing platform will need to be idealized. The baseline load paths for both bending and torsion load cases are then determined. The equations of equilibrium for each SSS will include the dimension parameters. It will therefore be possible to assess the effect of changing each dimensional parameter on the resulting forces, bearing in mind that the fundamental suspension input loads should also be changed if the vehicle's weight changes. The magnitude of the changes in loads and SSS edge forces can then be used to assess the risk of implementing the proposal.

8.2 Determine SSS outline idealization from basic vehicle dimensions

Before starting the idealization, the objectives must first be determined. The objectives will decide if the idealization should be local or global. More about that will be covered in section 8.3. For now it will be assumed that the objective is to idealize the *global*

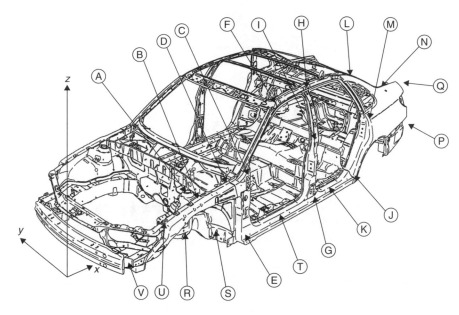

Figure 8.1 Structure of medium size saloon car (courtesy of General Motors).

load paths of an existing vehicle as shown in Figure 8.1. Dimensions critical to the idealization are the

- length from front suspension-to-body points to end of rear compartment pan;
- overall width of body floorpan between rocker sills;
- overall height of body from floorpan to roof;
- body width at suspension load points;
- fore–aft/vertical locations of the suspension load points;
- in-plane dimensions of the SSSs for dash, roof, sideframe, cowl, rear window, shelf, rear body opening, floor, windshield frame, and windshield inclination;
- location of internal members such as cross-bars and longitudinal beams.

The objectives for a global SSS idealization of an existing structure might be:

1. Assess impacts of a proposed change in basic vehicle dimensions.
2. Assess impacts of a proposed change in vehicle weight or payload
3. Assess impacts of a proposed change to how the suspension loads are introduced to the body (i.e. from a coil spring at the rear rail or a strut at the rear wheelhouse).
4. Aid in troubleshooting a particular problem that has occurred from testing, computer analysis, or in customer usage. This is most applicable on issues where a first order review of the basic load paths would add value to the problem solving process.

8.2.1 Locate suspension interfaces to body structure where weight bearing reactions occur

In the front of the vehicle it is the front strut tower, because that is where the springs support the front axle weight. In the rear it is the longitudinal frame rail. Note that on this vehicle the rear shock absorber forces are applied to the rear wheelhouse which may be considered part of the sideframe. Depending on the issue, it may be necessary to represent the shock and spring loads separately.

8.2.2   Generation of SSSs which simulate the basic structural layout

When first attempting to idealize the real-world structure it is tempting to represent too much detail. This usually results in a statically indeterminate representation. When more than one *significant* load path is possible, the structure should be idealized according to each, one at a time. That is, a separate model should be configured for each significant load path separately. This will result in a better understanding of the implications for each load path.

Attention should also be paid to the level of detail being represented. For example, if the vehicle structure rearward of the dash plane is of primary concern, then doing a detailed idealization of the front end structure will just add more complexity. It is the authors' general observation that if it takes more than 1 hour to idealize and generate the free body diagrams for a single vehicle structure, then probably more complexity is being represented than is necessary. Again, more will be said about the appropriate level of detail in section 8.3. Figure 8.2 conveys how the body structure example was idealized.

Once the idealization is established, free body diagrams are drawn to establish the force equilibrium equations. This will also uncover whether the idealization is statically determinate, which is a requirement for the SSS method.

The justification for the choice of the SSSs (12), (13), (14) representing the rear underfloor members can be appreciated from the view of the underfloor structure shown

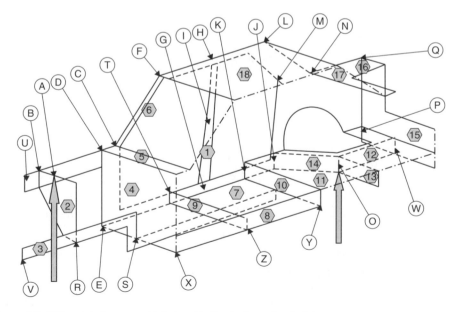

Figure 8.2 SSS model of structure (right-hand half only shown for clarity).

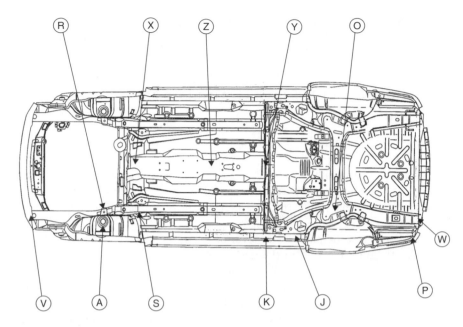

Figure 8.3 Underfloor structure of a medium size saloon car (courtesy of General Motors).

in Figure 8.3. The main beams form an H profile with the beams forward of the rear axle splayed out to join the rockers at node J (Figures 8.1, 8.2 and 8.3).

Once the idealization is established, free body diagrams are drawn to establish the force equilibrium equations. This will also uncover whether the idealization is statically determinate, which is a requirement for the SSS method. The next section describes the modelling in detail.

8.3   Initial idealization of an existing vehicle

Examination of the structure shown in Figures 8.1 and 8.3 reveals that there are many components that make up the passenger car body. One possible SSS idealization is shown in Figure 8.2 (only the right-hand side is shown for clarity). The corresponding nodes on the actual body and the SSS model are shown (corresponding nodes on right and left side of the body are referred with the same identification).

This SSS model consists of 18 SSSs but with the addition of corresponding members on the left-hand side that are not shown the complete model consists of 23 SSSs. The sideframe (1) is probably the largest single surface and includes the upper wing rail and the rear quarter panel. The front suspension tower (2) is represented by an SSS mounted transversely between the upper wing rail and the engine rail. Although the engine rail (3) is Z shaped it is still a plane surface and satisfies the principles of an SSS. The rear underfloor structure is represented by rear longitudinals (12), a rear cross-beam (13) plus an angled beam (14) connecting to the sill of the sideframe. This corresponds reasonably with the layout shown in Figure 8.3. The floor cross-beams (9) and (10) plus the rear panel (15) are connected to the two sideframes (1).

Table 8.1 Definition of nodes and coordinates of the body shown in Figures 8.1–8.3

Node ident.	Node description	X	Y	Z
A	Top of front suspension strut	900	560	800
B	Front fender longitudinal outboard of strut	900	750	800
C	Lower corner of windshield A-pillar	1450	750	800
D	Fender longitudinal to cowl	1350	750	800
E	Base of A-pillar/sill (rocker)	1450	750	170
F	Upper corner of windshield/cantrail	2050	750	1350
G	Base of B-pillar/sill (rocker)	2350	750	170
H	Top of B-pillar/cantrail	2550	750	1350
I	Middle of B-pillar	2450	750	760
J	C-pillar/sill (rocker)	3200	750	170
K	Floor crossbar (rear seat)/sill	2800	750	170
L	Upper corner of backlight/cantrail	3300	750	1350
M	Middle of C-pillar/front of parcel tray	3500	750	900
N	Rear corner of parcel tray/lower corner backlight	3850	750	900
O	Rear suspension spring mounting	3500	450	400
P	Rear lower corner of boot (trunk)	4400	750	400
Q	Rear upper corner of boot (trunk)	4400	750	900
R	Engine rail below A	900	500	400
S	Engine rail/dash panel	1350	500	170
T	Cross-beam (front seats)/sill	2000	750	170
U	Front end of upper fender rail	350	750	800
V	Front end of engine rail	0	500	420
W	Rear end of rear longitudinal rail	4400	450	400
X	Centre tunnel/dash	1350	0	170
Y	Centre tunnel/rear seat cross-beam	2800	0	170
Z	Front seat cross-beam/centre tunnel	2000	0	170

The upper parts of the structure are the cowl (5), the windscreen frame (6), the roof (18) and the backlight frame (17). The boot top (16) and the rear panel (15) are both horseshoe-shaped SSSs and probably show characteristics that are not ideal (see Figure 4.4(c)). Three further panels, the dash (4), the centre floor (7), the rear floor (11) plus the centre longitudinal beam (8) complete the SSS model structure. Figure 8.2 shows the front suspension support position at node A, and the rear suspension at node O.

After defining the node positions and the SSSs, before any calculations can be carried out, it is necessary to know the appropriate coordinates of each node. Using the coordinate system described in Figure 5.1 with the origin at approximately ground level, at the vehicle centreline and at the front bumper the coordinates of the nodes are obtained and listed in Table 8.1 for this vehicle.

8.4 Applied loads (bending case)

In Chapters 2 and 5 it was explained that for the bending case the mass of each major component is required and any dynamic factor appropriate to the particular vehicle. In this case study the static loads will be derived (they can be factored at a later stage) by simply multiplying the masses of components in kilograms by 10 to get the approximate forces applied to the structure in newtons. Using the factor 10 simplifies

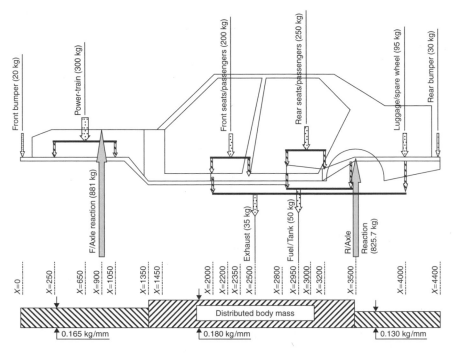

Figure 8.4 Typical load distribution for a medium size passenger car.

the arithmetic and the inaccuracy caused by the difference between 10 and 9.81 m/s^2 is insignificant.

Figure 8.4 shows typical masses, for a car of this size, of the major components, passengers, luggage and distributed body mass. The longitudinal positions of these masses are shown relative to the vehicle body. First, calculate the front and rear axle reactions, which is done as shown in Table 8.2. In the table the right-hand column shows the product of the mass and the distance from the front bumper. The total mass and the longitudinal position of the centre of gravity and hence the axle reactions are shown in Table 8.2 and Figure 8.4.

Finally it must be decided where the loads must be applied to the various SSSs (see Figure 8.5). In some cases such as the suspension reactions and bumpers the loads are applied to single nodes but for the power-train, passengers, fuel tank and exhaust these individual loads have to be distributed to appropriate positions. This is because nodes do not coincide with the centre of the masses. Care must be taken to ensure that the node loads are correctly proportioned and do not result in a longitudinal reposition of the resultant load. If this occurs then the axle reactions will be incorrect and it becomes impossible to maintain equilibrium conditions through the whole structure.

The power-train is often mounted on a subframe, which is mounted at two points on each engine rail. For this case it will be assumed these points are equally spaced 400 mm in front of and behind the power-train centre of gravity. The load on each point will be a quarter of the power-train, i.e. 750 N (75 kg). In addition the engine rails each carry half the load due to the front bumper mass.

Table 8.2 Calculation of axle loads

Component	Mass (M)	x-coordinate (mm)	Mx (kg-mm)
Front bumper	20 kg	0	0
Power-train	300 kg	650	195 000
Front pass/seats	200 kg	2200	440 000
Rear pass/seats	250 kg	3000	750 000
Fuel tank	50 kg	2950	147 500
Luggage/spare wheel	95 kg	4000	380 000
Rear bumper	30 kg	4400	132 000
Exhaust	35 kg	2500	87 500
Front structure	$(0.165*1350) = 222.75$ kg	$1350/2 = 675$	150 356
Pass compartment structure	$(0.180*2250) = 387$ kg	$(1350 + 3500)/2 = 2425$	938 475
Rear structure	$(0.13*900) = 117$ kg	$(3500 + 4400)/2 = 3950$	462 150
TOTALS	1706.75 kg		3 682 981

Longitudinal position of centre of gravity:

$$\bar{x} = 3\,682\,981/1706.75 = 2157.9 \text{ mm}$$

Moments about front axle:

$$\text{Rear axle load} = \frac{1706.75(2157.9 - 900)}{(3500 - 900)} = 825.7 \text{ kg}$$

Moments about rear axle:

$$\text{Front axle load} = \frac{1706.75(3500 - 2157.9)}{(3500 - 900)} = 881 \text{ kg}$$

Check equilibrium:

$$\text{Total axle load} = 1706.7 \text{ kg} = \text{Sum of Masses}$$

The front seats and passenger loads will be applied to the centre of the cross-beam at node Z, and to the sideframe at node T, the cross-beam to sill joint, and G the base of the B-pillars. Each seat and passenger has a mass of 100 kg and therefore loads of 214.25 N (21.425 kg) are applied to both nodes T and Z with a load of 571.5 N (57.15 kg) to node G. Node Z of course has a similar load for the second passenger on the other side of the vehicle.

Part of the weight of the exhaust will be applied at node Z, the centre of the cross-beam, while the remaining exhaust load will be applied equally to the two rear longitudinal rails between nodes O and W. As the exhaust centre of gravity is much nearer the front floor cross-beams the load distribution is 262.5 N (26.25 kg) at node Z and 43.75 N (4.375 kg) to each of the rear longitudinals. In practice the exhaust may be hung on one side only, which will lead to an unsymmetric load distribution. For simplicity make the model symmetric.

The rear seats and passenger loads are distributed to the rear cross-beam between nodes K and to the sideframe at node J. This means four equal loads of 625 N (62.5 kg).

The fuel tank is mounted under the rear seats and the loads from the tank are assumed to be applied at the same nodes as the rear passengers, that is to nodes K and nodes O. The proportions of the fuel tank loads are therefore 196.5 N (19.65 kg) each side of the cross-beam and 53.5 N (5.35 kg) to each node O.

The loads due to the luggage, spare wheel, and rear bumper masses are applied equally on each rear longitudinal rail. The loads due to the mass of the structure and

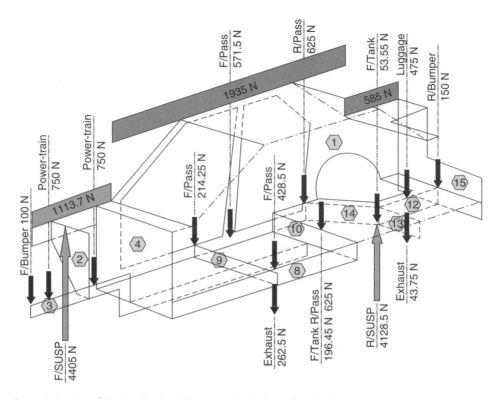

Figure 8.5 SSS model – bending loads (right-hand only shown for clarity).

its various fittings are assumed to be distributed uniformly over various parts of the sideframes.

The load condition on the right-hand side of the structure is illustrated in Figure 8.5. A complete picture of the overall loading condition of the structure has now been determined so the next stage is to determine the loads on each SSS.

8.4.1   Front suspension tower

The SSS representing the front suspension tower, surface (2) shown in Figure 8.6, is supported by the upper wing rail at B and the engine rail at R. This SSS acts as a beam supported at each end. Take moments about R to obtain:

$$K_B = \frac{4405(560 - 500)}{(750 - 500)} = 1057.2\,\text{N}$$

Resolve vertical forces:

$$K_R = 4405 - 1057.2 = 3347.8\,\text{N}$$

Figure 8.6 shows the suspension tower with forces, shear force and bending moment diagrams.

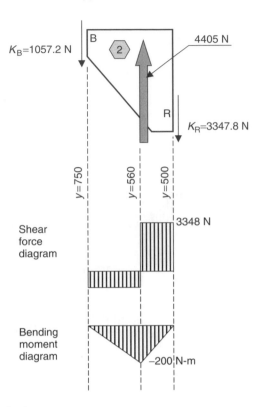

Figure 8.6 Front suspension tower.

8.4.2   Engine rail

This is represented by surface (3) in Figure 8.7 and the loading condition is as shown. Take moments about the rear end where K_{KK} acts onto the cross-beam under the rear seats and the value of K_S, the force into the dash panel, is obtained:

$$K_S = 1969.5 \, \text{N}$$

Resolve vertical forces:

$$K_{KK} = 221.7 \, \text{N}$$

Always check the result by taking moments about another point along the beam.

The maximum shear force on the engine rail is found to be 2498 N and the maximum bending moment (hogging) is 577 N-m. Both these maximums occur at the location of the suspension tower.

8.4.3   Centre floor

The load from the front passengers and seats has been assumed to act at the intersection of the front cross-beam and the centre longitudinal (Figure 8.8). Use the method described in the appendix to Chapter 4 to obtain the distribution of loads K_T, K_X and

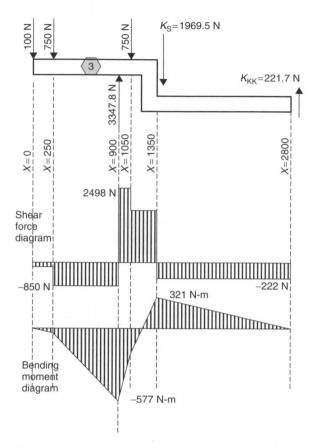

Figure 8.7 Engine rail.

K_Y. In using this method for this example it is assumed that the second moment of area of the centre longitudinal (SSS 8) is three times that of the cross-beam (SSS 9). The values obtained are:

$$K_X = 294.5 \, \text{N}$$

$$K_Y = 239.25 \, \text{N}$$

$$K_T = 78.62 \, \text{N}$$

The loads on these floor beams are oversimplified. The seat loads are not applied at the intersection of the cross-beam and the longitudinal but are applied at different points on the cross-beam and longitudinal. Similarly the reaction at the end of the engine rail may be considered to act on the front cross-beam. The equations for the deflection of the two beams with the real loads and load points that need to be equated then become more difficult to solve. However, the method used is considered satisfactory as a first approximation. With the model chosen, as may be expected, due to the greater stiffness of the centre longitudinal the reactions K_X and K_Y are much greater than the reactions between the cross-beam at the sills K_T. The bending moments and shear forces are much greater on the longitudinal but both members have sagging moments.

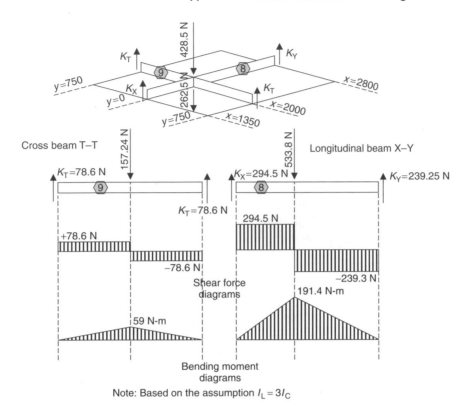

Figure 8.8 Floor beams.

8.4.4 Dash panel

Having determined both K_X and K_S the equilibrium condition for the dash panel (Figure 8.9) can now be determined. The edge loads K_{CE} are simply determined by resolving forces and by symmetry:

$$K_{CE} = 1822.25\,N$$

Note that the side parts of this panel are subject to the higher shear force and there is a hogging moment over the whole width of the panel.

8.4.5 Rear seat cross-beam

Figure 8.10 shows that there are a number of forces acting on SSS number 10. K_Y is the reaction from the centre longitudinal floor beam, then there are the loads from the engine rails K_{KK} plus loads from the rear passengers and the fuel tank. Resolve vertical forces to obtain that K_K, the forces between the beam ends and the sideframes, are 1162.8 N.

The shear force and bending moment diagrams show that this beam is subject to a sagging moment of 453.2 N-m at the vehicle centreline.

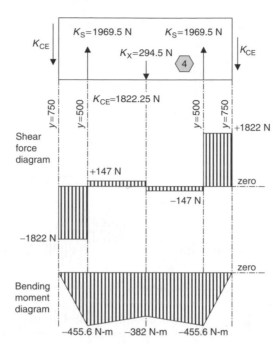

Figure 8.9 Dash panel.

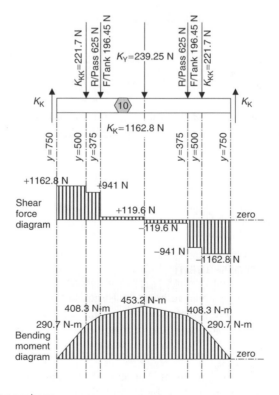

Figure 8.10 Rear seat cross-beam.

8.4.6 Rear floor beams

Underneath the rear floor there is a system of five beams (SSSs 12 (2 off), 13, 14 (2 off)). Loads applied to this system are from the luggage, spare wheel, exhaust, fuel tank and the reactions from the rear suspension (Figure 8.11). To determine the edge loads K_J and K_W take moments for the complete system about the rear end (the $x = 4400$ ordinate).

By this calculation obtain:

$$K_J = 2883.3 \, \text{N}$$

Taking moments about $x = 3200$ gives:

$$K_W = 672.9 \, \text{N}$$

Check the calculation by resolving vertical forces to ensure that there is no out-of-balance resultant force.

Examination of the shear force and bending moment diagrams shows that all beams are subject to hogging moments. The angled beams 14 have bending moments of -1223 N-m at node O while the longitudinal beams 12 have -865 N-m at the same node. The equilibrium of moments at this node is achieved by another bending moment in the cross-beam 13 also of -865 N-m. In this particular design the angle of the beams 14 to the vehicle x-axis is 45° so by resolving the bending moment in the angled beam

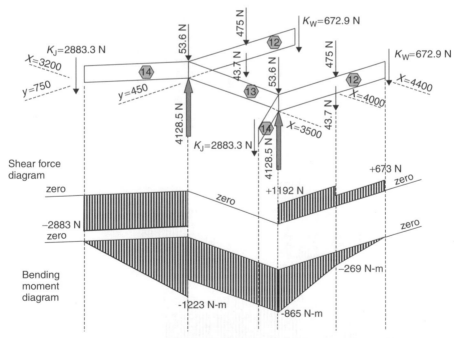

Note: Shear forces and bending moments for right longitudinal and left angled beam not shown for clarity.

Figure 8.11 Rear floor beams.

into longitudinal and transverse components it is found that:

$$-1223 \cos 45 = -1223 \sin 45 = -865 \, \text{N-m}$$

Therefore a satisfactory balance of moments is achieved.

Similarly by resolving vertical forces at node O then the two upward acting shear forces on the beams:

$$2883 + 1192 = 4075 \, \text{N}$$

are equal to the applied loads at this point:

$$4128.5 - 53.6 = 4074.9 \, \text{N} \quad \text{(a small rounding error of 0.1 N is acceptable)}$$

8.4.7   Rear panel

The remaining transverse SSS to be considered is the rear panel (Figure 8.12). In the previous paragraph the value of K_W was obtained. This force reacts against this rear panel and in addition the weight of the rear bumper is applied. By resolving vertical forces and by symmetry obtain:

$$K_{PQ} = 522.9 \, \text{N}$$

The shear force and bending moment diagrams show this panel has a constant hogging bending moment and zero shear over the centre section.

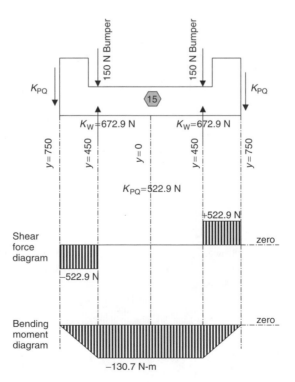

Figure 8.12 Rear panel.

8.4.8   Sideframe

From all the previous analysis the loads that have been determined are transferred out to the sideframe from the various SSSs. In addition, the side frame is assumed to carry the distributed loads of the body mass plus some additional loads applied directly from the passengers/seats. In Figure 8.13 we have the values of all the forces acting on the sideframe.

First, it is important to check that we have equilibrium by resolving vertical forces. For this example a small rounding error of less than 1 newton was found. Second, check that there is no resultant out-of-balance moment. Once again in this example anticlockwise moments about the front bumper were found to be $-14\,939$ N-m while the clockwise moments were $14\,939.9$ N-m which again is acceptable.

Finally plotting the shear force and bending moment diagrams shows high shear forces on the sideframe at the dash panel and where the rear floor beam is attached to the sill (node J). The maximum sagging bending moment of 937 N-m occurs near the B-pillar while the front top wing rail and rear quarter panels have smaller hogging moments.

8.4.9   Bending case design implications

Examination of the edge loads, shear force and bending moment diagrams indicates some of the design features that must be incorporated when designing the various components.

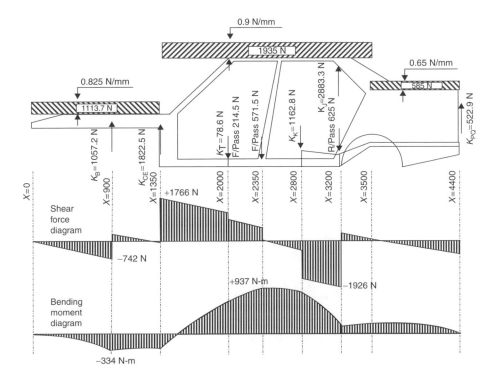

Figure 8.13 Sideframe.

The front suspension tower has high shear loads indicating that good shear connections must be made to the engine rail and to the upper fender rail. Because the tower is not very wide and it is very deep the bending loads are not of great significance.

In contrast the bending moment on the engine rail at the suspension tower connection is significant, as is the shear force connection to the dash panel. Alternative models could consider that the reaction at the rear of the engine rail be applied to the cross-member under the front seats. This would increase the shear forces to the dash panel and to the cross-member.

The floor cross-beam at the front seats and the central longitudinal beam have been treated in a simplified manner to determine the loads to the sills, the dash panel and the rear seat cross-beam. The seat loads could be distributed at four places on the cross-beam and two on the longitudinal, but this will lead to more complicated equations that are tedious to solve.

The bending moment on the dash panel is a hogging moment while that on the rear seat cross-beam is a sagging moment. As the dash panel is very deep bending is not a severe load condition but the side shear force on the deep panel may indicate the need of stiffeners (swaging). The rear seat cross-beam is not so deep and bending stresses may be significant. Good shear connections are required to the sills (rockers).

It should be noted that the rear floor beams although made of five SSSs have to be considered together in order to determine the shear forces and moments. The angled beam connecting the suspension mounting point to the sill carries the largest bending moment and shear force. This bending moment is balanced by moments in the rear longitudinal beam and in the short cross-beam. As the short cross-beam has a constant bending moment the shear force is zero and hence the suspension force is carried in shear to the angled and longitudinal beams. The need for a good shear connection to the sill and the need for a good joint connection of the three beams is indicated by these loads. Figure 8.3 shows a large pressing providing the connection between the beams which is necessary for carrying all the forces and moments generated at this point.

The rear panel is subject to a constant bending moment over the centre section where the depth is reduced for access to the boot. Fortunately, this moment is not excessive so for this case it does not cause any design difficulty.

Finally, examine the sideframe which has a large number of component loads, distributed loads due to the weight of the body itself plus the edge loads from the SSSs already mentioned. The maximum shear forces occur near the attachment to the dash panel and at the rear of the sill (rocker). In between these two sections there is a sagging moment with its maximum occurring near the B-pillar while the front and rear parts have hogging moments (i.e. the front fender and rear quarter panel act like cantilevers protruding from the rings around the door apertures).

8.5   Applied loads (torsion case)

In section 8.4 it was determined that for this vehicle the rear axle has the lighter load (see Figure 8.4). From Chapter 2 it was noted that the worst torsion load occurs when all the lighter axle load is applied to one wheel on that axle. This situation causes the other wheel on that axle to lift clear of the ground. The bending on the structure of course is still applied. To simplify the analysis the pure torsion condition is applied by applying equal and opposite wheel loads. The true situation can then be obtained by superimposing the results of the two cases.

Assuming that the track of this vehicle is 1.45 m, the pure torsion moment is:

$$4128.5\,\text{N} \times 1.45\,\text{m} = 5986\,\text{N-m}$$

This moment is applied to the rear structure at nodes O, which are 900 mm apart, therefore the loads applied to the structure at nodes O are:

$$R_{RT} = \frac{5986}{0.90} = 6651\,\text{N}$$

8.5.1   Rear floor beams

As in the bending case we consider the rear floor beams as a separate system shown in Figure 8.14. The applied loads R_{RT} must be transferred as shear into the three beams:

$$R_{RT} = P_J + P_W + P_{13}$$

The angled beams 14 have a shear force P_J acting at the sill attachment and therefore assuming the joint with the sill is a simple support, equilibrium is maintained by M_{14} where:

$$M_{14} = P_J{}^*(424.26) \quad 424.26 \text{ is the length of the angled beams.}$$

Similarly:

$$M_{12} = P_W{}^*(4400 - 3500)$$

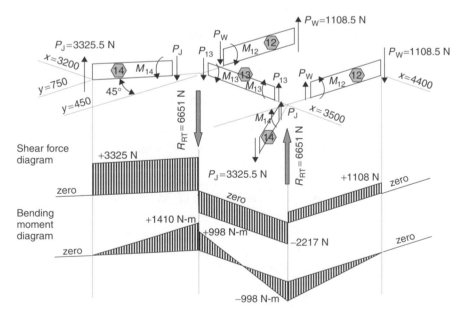

Note: Shear forces and bending moments for right longitudinal and for left angled beam not shown for clarity.

Figure 8.14 Rear floor beams.

Taking moments for the cross-beam 13:

$$2M_{13} - 900P_{13} = 0$$

As SSSs only carry moments in their own plane by resolving the components of the moment M_{14} in the plane of the rear longitudinals and in the plane of the cross-beam the following relationships are obtained:

$$M_{14}\cos 45 = M_{12}$$

$$M_{14}\sin 45 = M_{13}$$

Hence, from these six equations these values are obtained:

$$P_J = 3325.5\,\mathrm{N}$$

$$P_W = 1108.5\,\mathrm{N}$$

$$P_{13} = 2217\,\mathrm{N}$$

$$M_{12} = 998\,\mathrm{N\text{-}m}$$

$$M_{13} = 998\,\mathrm{N\text{-}m}$$

$$M_{14} = 1410\,\mathrm{N\text{-}m}$$

These values of shear forces and bending moments are illustrated in Figure 8.14. The right-hand angled beam (14) is subject to positive shear and a sagging bending moment, the left-hand rear longitudinal (12) has positive shear and a hogging moment. The beams on opposite sides will have opposite shear and bending moments. The short cross-beam (13) has a shear force but this beam is subject to contraflexure (i.e. hogging at left-hand side and sagging at the right).

8.5.2   Front suspension towers and engine rails

At the front of the vehicle, Figure 8.15, there is an equal and opposite torsion moment to that at the rear in order to maintain equilibrium. The mounting points of the suspension are 1120 mm apart, therefore:

$$R_{FT} = \frac{5986 \times 10^3}{1120} = 5345\,\mathrm{N}$$

The suspension towers (2) are short beams simply supported by the upper wing rail and the engine rail (3). The edge loads are obtained by resolving forces and taking moments:

$$P_B = 1283\,\mathrm{N}$$

$$P_R = 4062\,\mathrm{N}$$

The engine rail (3) reacts the load P_R and is supported by the dash panel (4), edge load P_S, and the rear seat cross-beam (10) with edge load P_{KK}.

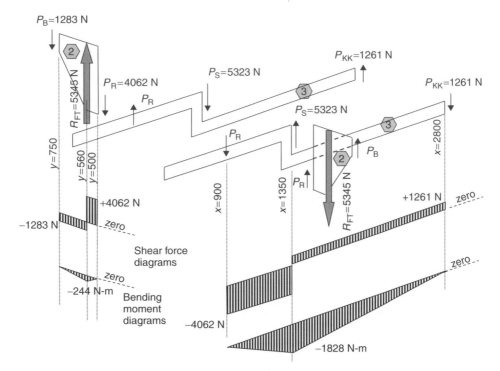

Note: Shear force and bending moments for left engine rail and right suspension tower only shown for clarity.

Figure 8.15 Front suspension towers and engine rails.

Resolving forces and taking moments gives the values:

$$P_S = 5323 \, N$$

$$P_{KK} = 1261 \, N$$

It should be noted that the loading system on the left side is the opposite of that on the right.

8.5.3   The main torsion box

In Chapter 5 it was shown how the passenger car structure acts like a torsion box. Figure 8.16 shows the torsion box structure for this vehicle. In the previous two sections the loads P_B, P_S, P_{KK}, P_J and P_W were evaluated and are shown with shaded arrows. Notice that they are of course equal but opposite in direction from the loads previously calculated (equal but opposite reactions).

First, consider the equilibrium of the dash panel (4). There is a moment applied from the engine rails by the forces P_S applying a clockwise moment as viewed from the front. Therefore, there must be a balancing anticlockwise moment by the edge loads Q_1 and Q_2. The moment equation is:

$$630Q_1 + 1500Q_2 - 1000P_S = 0 \tag{8.1}$$

The cowl (5) must react the edge load Q_1. This panel is held in equilibrium by complementary shear forces Q_3 and the moment equation is:

$$100Q_1 - 1500Q_3 = 0 \tag{8.2}$$

Similarly the windscreen frame (6) has complementary shear forces and the moment equation is:

$$813.94Q_1 - 1500Q_4 = 0 \tag{8.3}$$

Similarly for the roof (18):

$$1250Q_1 - 1500Q_5 = 0 \tag{8.4}$$

Similarly for the backlight frame (17):

$$710.63Q_1 - 1500Q_6 = 0 \tag{8.5}$$

Similarly for the boot (trunk) top frame (16):

$$550Q_1 - 1500Q_2 = 0 \tag{8.6}$$

The rear panel (15) has the forces P_W applied from the rear longitudinal beams, therefore consider this in a similar way to the dash panel. The applied moment from forces P_W is anticlockwise as viewed from the front:

$$500Q_1 + 1500Q_8 - 900P_W = 0 \tag{8.7}$$

The rear floor (11) is in complementary shear, therefore:

$$1600Q_1 - 1500Q_9 = 0 \tag{8.8}$$

The cross-beam (10) under the rear seats is loaded in a similar manner to the rear panel, therefore:

$$230Q_1 + 1500Q_{10} - 1000P_{KK} = 0 \tag{8.9}$$

Completing the periphery of the shear box is the centre floor (7), which is loaded by complementary shear forces:

$$1450Q_1 - 1500Q_{11} = 0 \tag{8.10}$$

Now check that the correct direction of the shear forces Q_1 has been maintained by comparing the sense of the arrows between the front of the floor and the lower edge of the dash panel.

Finally, consider the right-hand sideframe. The loads on this member are the equal and opposite reactions to edge loads Q_2 to Q_{11} plus the forces P_B and P_J already evaluated. The left-hand sideframe will be loaded with exactly equal but opposite loads. Consider moments about the lower corner of the windscreen, i.e. the point through which Q_3 and Q_4 act. This eliminates two terms in the equation and simplifies the algebra/arithmetic:

$$550P_B + 100Q_2 - 550Q_5 - 2400Q_6 \cos \beta - 100Q_6 \sin \beta - 100Q_7 + 2950Q_8$$
$$- 400Q_9 + 1350Q_{10} - 630Q_{11} + 1750P_J = 0 \tag{8.11}$$

Table 8.3 Torsion Case Edge Loads

Edge load	Initial model force (N)	Initial model shear flow (N/mm)	Alternative model force (N)	Alternative model shear flow (N/mm)
Q_1	2771	1.85	3440	2.29
Q_2	2385	3.76*	1263	2.00*
Q_3	185	1.85	1705	170.49*
Q_4	1504	1.85	1866	2.29
Q_5	2309	1.85	2866	2.29
Q_6	1313	1.85	1629	2.29
Q_7	1016	1.85	–	–
Q_8	−259	−0.52*	–	–
Q_9	2956	1.85	–	–
Q_{10}	416	1.81*	527	2.29
Q_{11}	2679	1.85	1391	0.96*
Q_{12}	–	–	802	2.29
Q_{13}	–	–	1337	2.29
Q_{14}	–	–	917	2.29

*Not complementary shear

There are therefore 11 unknowns Q_1 to Q_{11} and 11 equations (8.1 to 8.11), which can be solved by matrix algebra or by commercially available software. Using Maple V software the values obtained are as shown in Table 8.3.

Having obtained all these values confirm these results by resolving forces vertically and horizontally on the sideframe. To ensure equilibrium is obtained resultant forces in each direction must be zero.

Resolving forces vertically (positive upward):

$$+4855.3 - 4757.8 = 97.5\,\text{N}$$

This means an out-of-balance force of 97.5 N, which is 2.0 per cent of the total positive force, which is acceptable.

Resolving forces horizontally (positive to the rear):

$$+5635 - 5547.1 = 87.9\,\text{N}$$

This shows an error of 1.6 per cent, which again is acceptable.

Another check is the shear flow (shear force per unit length) on each of the SSSs subject to complementary shear. These should be equal (see Table 8.3). Panels that do not have complementary shear will have different shear flows.

8.5.4   Torsion case design implications

The values of the edge loads calculated for the rear floor beams (Figure 8.14) illustrate the high forces and moments that occur at the rear suspension mounting points (node O). The values of the shear forces and bending moments are of different character and higher values than for the bending case. These forces and moments indicate why this car has large sections for these beams as illustrated in Figure 8.3.

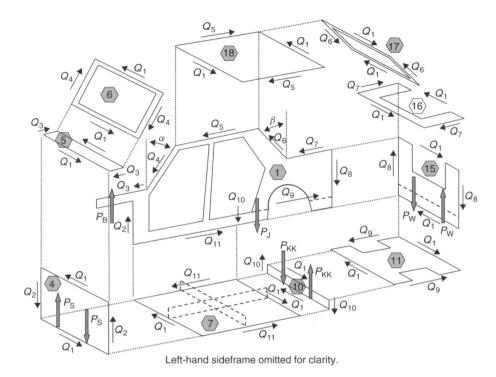

Left-hand sideframe omitted for clarity.

Figure 8.16 The main torsion box.

The loading on the engine rails shows much higher shear forces and bending moments (Figure 8.15) than occurred in the bending case again indicating the requirement for large beam sections.

Examination of the shear forces on the main torsion box (Figure 8.16) shows the importance of maintaining a continuous shear path over the windscreen frame, roof, backlight frame, boot, floor and dash panel. Components that are likely to cause design difficulties are the windscreen frame, the backlight frame, the boot top and rear panel. To maintain strength and stiffness the windscreen and backlight frames must have adequate section stiffness and good corner joints where the maximum bending occurs (see Figure 7.15). The boot top frame and rear panel with the large cut-outs can result in flexibility. To reduce this the sides of the boot top can be kept wide and the rear edge of the rear panel raised. Narrowing the boot lid and raising the edge of the rear panel reduce the accessibility to the boot. The rear floor with the cut-outs for the rear wheels is usually provided with sufficient stiffness by the wheel arches.

8.6   An alternative model

The modern integral structure of a passenger car is a highly redundant structure, but in this SSS model the structure has been simplified to be statically determinate. Effectively some structural members have been removed or their effect ignored. Therefore, there are different ways the structure can be modelled (i.e. different SSSs can be chosen).

For this existing vehicle consider the alternative front structure shown in Figure 8.17. This model distributes the inboard load from the suspension tower by shear to the inner wing panel (3b). The panel is supported by the top rail (3c), the engine rail (3a) and the dash panel. The top rail is then supported by the cowl and the engine rail (3a) by the centre floor (7). The loads in the inner wing panel, the top rail and the engine rail are shown in Figure 8.17.

The main shear box is modelled with a structural ring (19) behind the rear seat (Figure 8.18). From the view of the structure shown in Figure 8.1 it can be seen there are large gussets between the floor, rear parcel tray (20) and sideframe effectively making a ring structure. This, with the rear parcel tray, provides an alternative shear path between the backlight frame and the floor.

To reduce the length of this chapter only the torsion case is examined for this alternative model. The reader may wish to consider the bending case.

8.6.1   Front suspension towers and inner wing panels

Starting with the system shown in Figure 8.17 the load P_R is found to react into the inner wing panel. By resolving forces vertically P_S is equal to P_R and by taking moments the end loads P_{CC} and P_{EE} in the top rail and engine rail are found to be 2901 N. Resolving forces horizontally, these must be equal and opposite. The forces

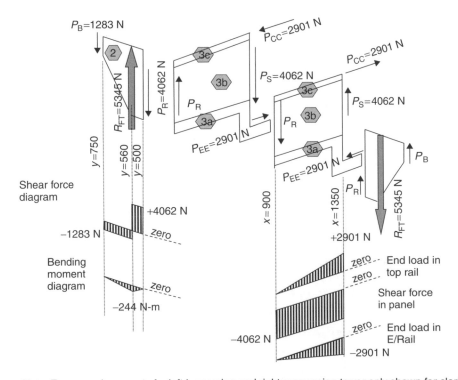

Note: Forces and moments for left inner wing and right suspension tower only shown for clarity.

Figure 8.17 Front suspension towers and inner wing (alternative model).

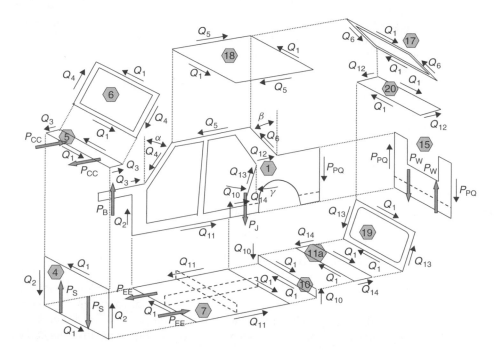

Figure 8.18 The main torsion box (alternative model).

P_{CC} and P_{EE} are reacted into the cowl and the floor panel, both of which are components of the main torsion box (Figure 8.18).

8.6.2   Rear floor beams

The loading on these beams remains exactly the same as for the previous model as shown in Figure 8.14.

8.6.3   The main torsion box

The loads applied to the main torsion box from the front suspension tower, inner wing panels and rear floor beams are shown with large arrows in Figure 8.18. The dash panel (4) is loaded with forces P_S, the cowl (5) with forces P_{CC}, the floor (7) with forces P_{EE}, the rear panel (15) with forces P_W and the sideframe (1) with P_B and P_J. The rear panel is now not part of the main shear box but a beam that transfers loads P_{PQ} to the sideframe:

$$P_{PQ} = P_W 900/1500 = 665.1\,\text{N}$$

The moment equations for the various SSSs of the torsion box are now:

Dash panel (4)

$$630Q_1 + 1500Q_2 - 4\,062\,000 = 0 \tag{8.12}$$

Cowl (5)

$$100Q_1 + 1500Q_3 - 2\,901\,400 = 0 \tag{8.13}$$

Note this is now *not* in complementary shear and that the directions of forces Q_3 have been changed.

Windscreen frame (6)

$$813.94Q_1 - 1500Q_4 = 0 \tag{8.14}$$

Roof (18)

$$1250Q_1 - 1500Q_5 = 0 \tag{8.15}$$

Backlight frame (17)

$$710.63Q_1 - 1500Q_6 = 0 \tag{8.16}$$

Rear parcel tray (20)

$$350Q_1 - 1500Q_{12} = 0 \tag{8.17}$$

Rear seat ring frame (19)

$$583.1Q_1 - 1500Q_{13} = 0 \tag{8.18}$$

Rear seat panel (11a) (this is now only the front part of the rear floor)

$$400Q_1 - 1500Q_{14} = 0 \tag{8.19}$$

Rear seat cross-beam (10)

$$230Q_1 - 1500Q_{10} = 0 \tag{8.20}$$

Note this beam is now loaded with complementary shear and the directions of forces Q_{10} have been changed.

Floor panel (7)

$$1450Q_1 - 1500Q_{11} - 2\,901\,400 = 0 \tag{8.21}$$

This panel is now not loaded with complementary shear because of the forces P_{EE}. It could be assumed that the direction of Q_{11} is changed but it has been left unchanged.

Sideframe (1)

$$550P_B + 100Q_2 - 550Q_5 - 2400Q_6 \cos\beta - 100Q_6 \sin\beta + 100Q_{12}$$
$$- 2050Q_{13} \cos\gamma + 100Q_{13} \sin\gamma - 400Q_{14} - 1350Q_{10} - 630Q_{11} + 1750P_J$$
$$+ 2950P_{PQ} = 0 \tag{8.22}$$

Solving these 11 equations (8.12 to 8.22) using the Maple V software the values of the edge loads obtained are shown tabulated in Table 8.3.

Check the results to ensure equilibrium of the sideframe by resolving vertical forces. The total upward forces are 4855 N and downward 4758 N.

$$\text{Rounding error} = 4855 - 4758 = 97\,\text{N} \quad (2.0\% \text{ error})$$

Resolving horizontal forces, forces to the rear are 5635 N, and forces towards the front 5547 N.

$$\text{Rounding error} = 5635 - 5547 = 88\,\text{N} \quad (1.6\% \text{ error})$$

The final check is to evaluate the shear flow to ensure that panels subject to complementary shear all have the same shear flow. These are tabulated in Table 8.3 and found correct.

8.6.4   Torsion case (alternative model) design implications

At the front end this model requires a good shear connection between the inner wing panel and the dash panel and a stiffening member between the top of the suspension tower and the cowl. The cowl for this model carries a significant bending moment in addition to the shear force transferred between the dash panel and the windshield frame. Attachment of the engine rail to the central floor must ensure the end load P_{EE} is distributed as shear into the floor. There will be local bending at the Z shape of the engine rail adjacent to the dash panel but theoretically no bending occurs on the engine rail as it extends under the floor.

At the rear end the rear panel now acts as a beam between the two sideframes and is not required to carry shear between the floor and boot top frame. Similarly, the boot top frame does not carry shear between the rear panel and the backlight frame. Hence, the less effective SSS members with large cutouts are not required to be part of the shear box. Instead, the shear box is completed by the rear parcel tray and the ring frame behind the rear seats. Provided this has large corner gussets, which this vehicle has, there is a suitable path for transferring shear from the backlight frame to the floor. The rear seat cross-beam is simply a panel subject to complementary shear due to no loads being transferred from the rear end of the engine rails. Note that in both models of the structure the centre longitudinal and the cross-beam under the front seats do not contribute to the torsion load case.

Table 8.3 shows that for the alternative model the shear flow values are higher than in the original model. This is because the enclosed area of the main torsion box has been reduced. If we take a section through the body at the vehicle longitudinal centreline the original model includes the boot (trunk) area while the alternative model excludes this area. From theory of thin walled torsion boxes the shear flow $q = T/2A$ where T is the torque applied and A the enclosed area. The alternative model has a reduced enclosed area and therefore the results of the models agree with this theory.

8.7   Combined bending and torsion

In practice the pure torsion load cannot be applied alone due to the fact that gravity is present, and bending will always occur. To find the effect of static bending and torsion simply add the loads that occur due to the two cases. For example, add the load conditions illustrated in Figure 8.11 and 8.14. Also the torsion loads can all be reversed as the vehicle can traverse many different surfaces and be twisted in the opposite direction to that analysed. Therefore the maximum shear at the rear suspension mounting (node O) in the rear longitudinal beam is 2300 N and maximum bending moment 1863 N-m. For the angled beams the maximum shear is 6108 N and maximum bending 2633 N-m while for the cross-beam the maximum bending is 1863 N-m.

Other components of the structure can be considered in a similar way. Dynamic or load factors will also need to be considered.

8.8 Competing load paths

From this worked example it will be realized that there are different SSS models and hence alternative load paths. Therefore the questions that arise are – 'Which model is correct?' or 'Which model is more accurate?' With the SSS method these questions cannot be answered and illustrate the limitations of this method. Nevertheless the SSS method has illustrated the type of loads that are applied to each SSS or major component. In practice both of these models are partially correct. The shear force path between the backlight frame and the floor is via both the boot/rear panel and the seat back ring frame. At the front the engine rail will carry both end loads as shear into the centre floor and also bending moments that are reacted by loads into the dash panel and the cross-beams. The ratio of sharing the loads by the different load paths/methods cannot be determined unless more complex analysis such as finite element methods are applied.

In spite of the limitations of the SSS method the analysis has provided a view of the possible load paths through the structure. It has indicated where high loads are to be found such as the engine rail near the dash panel and on the rear floor beams. The difficulty of maintaining a suitable shear path around the boot and rear seat has been revealed. The vehicle designer now has the characteristics and nominal values of the loads on the major structural components and can use this information to ensure the required structural features are incorporated in the design.

9

Introduction to vehicle structure preliminary design SSS method

9.1   Design synthesis vs analysis

The differences between synthesis and analysis are explained by an example problem from an engineering textbook: The first problem is a beam with a combination of uniformly distributed and concentrated loads. The question is asked – 'What section size is required to support these loads?' Shear and bending moment diagrams determine the maximum moment and shear forces, from which basic equations are used to calculate the required minimum moment of inertia that the cross-section must be designed for. A variety of different 'answers' can provide what shape and thickness the beam should take, given the design constraints. Because the beam must satisfy both yield and deflection requirements, calculations using both WL^3/EI and Mc/I equations are used. Two inertia properties are calculated from these and a selection is made which will satisfy both requirements. From this information it is also determined whether it is a stiffness or stress governed design. Shape, thickness and material properties are then proposed based on this data. An optimization for minimum weight may be performed to arrive at a final cross-section requirement. What has just been illustrated, albeit with a very simple example, is *synthesis*.

Now compare this to the second problem: Here the beam type, length and loads are the same as the first problem. The exception is that the cross-section shape, size and thickness have already been selected by the designer before any calculations have been performed. The question is asked – 'Can this structure carry the loads?' The same equations are used, but this time to *analyse* the given beam's structural adequacy.

It may be thought that in the first problem, the function of the beam (to carry a set of given loads) was considered early in the design process before the cross-section was designed. The final cross-sectional design reflected these requirements. In the second problem where *analysis* was applied, the cross-sectional design was defined before it was determined whether it would support its function. If this question is asked at a much later stage after packaging and other considerations have already defined the beam's parameters, then the results may cause time-consuming design rework. This approach may also require more iterations to reach the target and limit opportunities for optimization.

9.2   Brief outline of the preliminary or conceptual design stage

This section may be described as 'the need for speed' in developing new designs, because the nature of the market today is two-fold.

(a) Products must be refreshed within a shorter product life cycle (between the old and new model introduction time span) in order to maintain customer enthusiasm and to address shifting consumer tastes.
(b) Once new niche markets are identified, the first product to get there tends to reap the greatest reward. It is therefore important that the design process be not only fast, but smart.

Consequently it is desired to have initial data upon which to base early structural design studies. This need stems from two motivations:

(a) The utilization of preliminary finite element models to help develop requirements.
(b) The utilization of the preliminary models to evaluate and confirm concepts against requirements once design data are more developed.

At the beginning of the concept design phase, the available data often change quickly over a fairly short period of time as packaging and other issues are reconciled and requirements are debated. Many design alternatives may be under consideration. Further, the design process does not necessarily stop in order to wait for someone to model the structural concept and give feedback before the design changes. The design is in a constant state of flux. For the engineer operating in the analysis mode, it is analogous to trying to catch a high speed moving train. Time can be saved by applying synthesis methods up front to direct design and downstream analysis efforts. Effective synthesis is dependent first on having good requirements to work from. These are not just requirements in terms of a global structure specification, such as a first natural frequency or torsional stiffness, but targets and guidelines that a designer can utilize to create geometry. It is here that the SSS method can have a significant impact, both prior to and during conceptual FEA. This is because the SSS method has its root in understanding the fundamental structural principles employed in the concept.

In the beginning stage of a new vehicle design, there are usually not enough data available to assemble a complete FE model of the *new* structure. Models that can be assembled at this early stage may be coarse, such as beam 'stick' type and/or hybrids of coarse and detailed mesh adapted from current or past designs. These models are typically used for synthesis work. The SAE literature cites numerous articles of coarse FE models being applied up front to automotive body structure design. Among those is the paper by Longo, *et al.* (1997).

The SSS method used prior to and during the early design stage can complement computer aided synthesis. This will be further explained in section 9.4. Once the design is underway and iterations are being performed, the SSS method can provide guidance in interpreting FEA results and in developing proposals for FEA to evaluate. FEA results are usually expressed in terms of static deflections and internal reaction forces on elements that can be used to assess whether a desired load path, qualitatively developed by SSS models, has been achieved.

9.3   Basic principles of the SSS design synthesis approach

This section will illustrate how SSS can be applied at each stage of the design synthesis process and how its application can help work the issues inherent with vehicle design change proposals.

9.3.1   Starting point (package and part requirements)

The SSS method can be *qualitatively* applied for load-path visualization when only limited dimensional information about the vehicle body is available. The examination of load paths for various body style types has been shown in previous chapters. The equations for the edge forces and internal reactions show that the method requires only the very basic parameters of the body be known or estimated. Among them are: length, width, height, external load inputs, front/rear axle weight and distribution, and lumped masses of major subsystems. This enables the SSS method to become part of the design synthesis process during package studies and during the development of design requirements. This is illustrated with the following example.

It is proposed to make two different size vehicles off the same platform. That is, it is initially envisioned that the floor panel can be widened and lengthened while extending the sill (rocker). It is proposed that the rocker sill cross-section be common. The dash panel shall be widened accordingly, but it is proposed to keep the motor compartment side panel structure common, with only a width change to the upper and lower front cross-members. Similar architectural changes are envisioned for the rear structure. It is proposed that such an arrangement might save considerable design, development and tooling costs. Performance requirements may be somewhat vague, for there are at least several options. To realize the desired programme cost savings, two alternative design approaches might be:

(a) The frame member sections must be designed so that the bending and torsional stiffness of the larger vehicle is not compromised. In essence, the smaller vehicle is allowed to be overdesigned.
(b) The frame members are envisioned to be based on the smaller vehicle. Any differences in the larger vehicle are to be accommodated by additional reinforcements only, which in turn must be compatible with the manufacturing process.

While well-documented requirement paragraphs may be provided at the vehicle level, it is difficult with a new design to know very early if either strategy will satisfy the mass and cost constraints of the total vehicle programme. Not surprisingly, a current or previous vehicle with the desired performance characteristics may be chosen for the initial package study. While this may result in less risk, it may also negate opportunities to look at alternative designs that might ultimately lead to lower cost, and a greater success in the market place.

Returning to our example, basic engineering fundamentals suggest that the larger vehicle will have less bending stiffness than the smaller vehicle. This is a consequence of the basic beam equation: $d \propto WL^3/EI$. A bending moment diagram (where the beam is assumed to be a SSS) might show qualitatively where reinforcements may need to be incorporated into the rocker sill members. Alternatively, it may show where increased upper structure stiffness may be needed, resulting in stiffer pillar and roof structures for the larger vehicle, or a combination of both. This is somewhat analogous to a road bridge structure whose bending stiffness at mid-span is dependent on the truss framework.

The following is a review of the example using the front motor compartment side panel and dash structure. The larger vehicle has both a wider dash and, because of its mass, higher input loads from the suspension. Reviewing the equations for the dash under vehicle torsion (Chapter 5, section 5.3.4) will indicate the sensitivity of the edge forces to its width and height dimensions. The higher mass will produce proportionately larger edge forces due to the higher input loads. These in turn can effect the requirements for the adjacent structural members as well, depending on their governing load case.

The front floor pan cross-member at the tunnel finds it must react higher loads due to the greater vertical load applied at the front suspension points during the fundamental bending load case described in Chapter 8, section 8.4. This suggests the need for additional structure that may not have been considered originally. The need for the front motor compartment side panel structure to carry higher loads must also be considered. Additional reinforcements and/or increased section dimensions may be indicated.

The previous examples just cited briefly showed how basic dimensional and weight information together with the SSS method and basic fundamentals can help provide direction and identify potential areas of concern early in the process. Similar thought processes can be used for understanding changes required for, say, a taller vehicle based on an existing shorter one. Recall that the equilibrium equations have dimensional parameters which function as coefficients for the matrixes which are explained in previous sections of this book. An examination of the equilibrium equations solved for each SSS can yield useful information about the sensitivity of basic dimensional changes on the resulting forces and moments.

Aesthetic design or styling also has a significant influence on the body structure. The interior and exterior design themes are intended to send a message to the target customer that will evoke favourable reaction and perception about the vehicle. Once enough information has been pulled together, the first priority is to develop a balance between packaging (vehicle layout), styling, and function. Packaging determines how the occupants will sit and how the power-train and chassis are to be placed. Function relates to the various requirements that the vehicle must perform, such as (a) resist noise and vibration, (b) improve ride and handling, (c) maintain durability, and (d) manage crashworthiness etc. Styling emphasizes the visual design features that are needed to distinguish the package from competitive vehicles. Early decisions about the target market, occupant accommodation, weight limits, power-trains, chassis systems, and energy absorbing crush space tend to drive the basic width, length, and height of the vehicle.

The result of this iterative and interactive balancing is an envelope within which the engineer will optimize the body structure. It is imperative that the structural engineer be involved during this phase. There will be many issues where proactive input is essential. To the extent that decisions tend to be based upon where the consequences are the most easily understood and visibly tangible, it is important for the structural engineer be able to communicate. Packaging studies, for example, illustrate interference issues that are relatively easy to visualize and comprehend. That is, two objects cannot occupy the same space. There might be negotiation on clearances, but a choice may need to be made between redesigning a structural member or widening the vehicle. Widening the vehicle may not be a good solution if it is inconsistent with the target market size and weight constraints. Redesigning the structural member may violate the business case because it would require new tools and additional parts resulting in increased cost. The challenge is to develop a body structure design that is inherently less sensitive to packaging issues while meeting global body performance requirements and still be within size, manufacturing, cost, and mass constraints. This is especially challenging for smaller vehicles with multiple power-train and chassis combinations that are needed to satisfy global markets.

Crashworthiness issues are also a significant driver. Crash performance requirements affect the structural member location, size/shape/thickness, material selection, and chassis/power-train mounting support configurations. It is outside the scope of this book to cover crashworthiness fundamentals.

Ground clearance lines, car wash rack clearances, bumper heights, and ramp approach angles can affect the packaging of the floor pan and frame rail structure.

The myriad constraints and issues make it essential that alternative body structure designs be considered upfront. The one which worked for a current or previous model may require for the new vehicle that significant issues be addressed in order to package the newly desired power-train.

9.3.2   Suggested steps

During the preliminary design phase, it is recommended to start first with the bending and torsion load cases with SSSs at the total body structure level instead of the body subunit level. The suggested steps are as detailed in Chapter 8, section 8.4, and Chapter 8, section 8.5:

1. Estimate the loads and loading conditions, as in Chapter 3 and Chapter 10, section 10.4 or from simulation.
2. Draw free body and loading diagrams.
3. Calculate the internal, non-edge reaction forces.
4. Construct shear and bending moment diagrams.
5. Formulate the matrix equations to solve for the edge forces.

9.3.3   Suggested priorities for examination of local subunits and components

Steps1 to 4 should be repeated for local subunits and components. The load cases from Chapter 2, section 2.4 should be walked through to identify the critical or governing load conditions. Certain simultaneous load combinations, such as in Chapter 2, section 2.5, should then be examined. Where possible, simulation and/or test data should be utilized to validate the load factors selected and the assumptions concerning simultaneous load conditions. The order for examining local subunits and components is suggested below:

1. Lower structure (floorpan, motor compartment, rear compartment panel, rocker sills), including chassis and power-train attachment points.
2. Dash and rear seat back panel, or rear seat back opening.
3. 'A'-pillar (windshield and front hinge pillar).
4. 'B'-, 'C'-, and 'D'-pillars.
5. Cantrails (roof side rails or door headers), windshield and backlight glass, or rear hatchback lid headers.

9.3.4   Positioning of major members

The SSS method lends itself to quickly examining alternative load paths in a qualitative sense which, when combined with FEA, will yield recommendations on the placement of major structural members. Alternatives might be the trade-offs between the floor member or the sideframe as the primary suspension load reaction point on the body. Another might be the trade-off of reacting the lower front motor compartment rail vertical loads with a torque box, or with the dash and front seat cross-member. It is important that all the relevant requirements be considered, including those which the SSS method does not examine, such as crashworthiness. Once the packaging balance has been achieved, there may be limited opportunity for further alternative structural members or sizing changes. Material specification, gauge thickness, and internal reinforcements may still be altered at this stage. It is, therefore, important that the body structure engineer be actively engaged while the packaging is being developed so that alternatives and optimum sizing can be considered.

9.3.5   Member sizing

The loads derived from the SSS method can be used with fundamental engineering mechanics for preliminary sizing of members which are governed by elastic stress criteria (no yield or permanent deformation) under the loading conditions described in Chapter 2, section 2.4. Here it is assumed that these are the governing load conditions. This may or may not be the case. Some members may be governed by crashworthiness requirements. Others may be governed by stiffness, noise and vibration, or ride and handling requirements. It is important for the engineer to identify the governing load cases for each major structural member. Finite element methods are especially applicable when the structure is statically indeterminate, which, as stated earlier, is a characteristic of modern unit body construction. There are illustrative examples showing how preliminary sizing for certain load cases was performed on unit body structures before finite element analysis was common practice. An example of this process was provided by T.K. Garrett (1953). Starting from loading, shear and bending moment diagrams, Mr Garrett walked the reader through methods for determining the size of the rocker sill and also suspension interface attachment points. Assumptions were made on the load-path distribution that resulted in reasonable sizing of the members for the fundamental bending load case. Had FEA been applied then, it would have been possible to further optimize the design solutions proposed. Mr Garrett's article is an example of early design synthesis.

9.4　Relation of SSS to FEA in preliminary design

It is important that the scope, limitations, and suggested role of SSS in the design process be reviewed if it is to be applied effectively and with other tools:

9.4.1　Scope of SSS method

- Fundamental bending, torsion, vertical, lateral and fore–aft loads on the body.
- Total body-in-white structure, subunits, and components of the structure.
- Unit body construction. (It is conceivable that a body structure that is bolted to a chassis frame could be assessed using the SSS method, provided it was examined separately from the chassis. In this case, the input loads would be from the body mount locations.)

9.4.2　Limitations and assumptions of SSS method

- Statically *determinate* solutions for each alternative load path.
- *Qualitative* recommendations and assessments for *stiffness* governed members.
- *Quantitative preliminary sizing* and recommendations for *elastic stress* governed members.
- *Static* equivalent simplification of dynamic load cases.

9.4.3　Suggested role of SSS method

- Organize the thought process for visualizing and rationalizing the fundamental loads (mentioned above) going into the body structure.
- Facilitate a disciplined thought process for qualitative synthesis of the body structure.
- Be a starting point for preliminary structural sizing before FEA is applied, subject to the limitations and assumptions described above.
- In conjunction with FEA, serve as:
  - a means to help explain and interpret FEA results;
  - a means of generating ideas for FEA iterations.
- Help establish a load-path philosophy for the structural design execution that will carry forward through subsequent design stages.

9.4.4　Role of FEA

- *The* primary tool for validating design before hardware.
- Solve for statically *indeterminate* conditions.
- Extract internal loads from which detailed structural sizing can occur.
- Means of providing quantitative synthesis results.
- Analysis of design at appropriate stages to confirm or provide specific direction.
- Structural optimization.

9.4.5   Integration of SSS, FEA and other analyses

We have discussed briefly the roles of both SSS and FEA, along with the potential benefits of using the tools together in process. It is suggested that the use of these techniques side by side, rather than separate independent application, will have the highest potential value. The understanding of SSS in the hands of a qualified FEA engineer could potentially lead to smarter, timelier iterations needed to develop solutions for the applicable loading conditions. Other types of analyses that may potentially be used in close cooperation with the SSS method are dynamic models that simulate and predict loads on the vehicle structure.

The data from a typical body conceptual finite element analysis may consist of deflections, natural frequencies and perhaps element stress output. Deflections are typically animated to amplify and illuminate structural deficiencies such as discontinuities and hinge points. Certain deflections may be computed to provide global bending and torsional stiffness values that are typically based on the experience of the company's successful models and competitive vehicles. A normal modes solution will yield natural frequencies and mode shapes. The mode shapes are typically animated to amplify and illuminate structural deficiencies such as discontinuities and hinge points. The eigenvalues will be compared to a first natural frequency target that is again typically based on the experience of the company's successful models and competitive vehicles. Strain energy and strain energy density contours are typically plotted to highlight potential problem areas for each mode of interest. There is much literature about finite element applications that will not be repeated here. The point of this section is to delve briefly into how the FEA could potentially work hand in hand with the SSS method. The example below will be a hypothetical problem solving scenario.

Example of FEA working in conjunction with the SSS method

A finite element analysis of a conceptual station wagon or estate car body structure has revealed that the first structure mode natural frequency is below target. This first structure mode happens to be a global torsion mode. Animations show considerable match-boxing or lozenging of the rear hatch door frame opening. Strain energy plots indicate relatively high concentrations at the upper corner joints of the frame as well. Model checks reveal no problems with element connectivity. Before further iterations are performed, free body diagrams of the estate car are drawn for the torsional load case. The diagrams show that the rear hatch door frame must be capable of carrying shear flow edge forces in order for the structure to be in equilibrium, and that the maximum bending moment occurs in the corners. It is evident from the animation that the frame is hinging or lozenging at the corners in the finite element model. It is also noted that the hatch frame is not entirely continuous, in that it has a step or transition at the waist (beltline) to accommodate a desired styling feature. This relates back to the free body diagram in Chapter 6, section 6.3 which indicated that this discontinuity fundamentally induces internal forces on the frame that challenge the shear flow or edge forces. The interruption of the smooth shear flow implies that the frame will have to carry more of the load in bending, and that perpendicular forces will be introduced to adjacent SSSs. Further it is noticed that the corners of the frame have large access holes to accommodate welding and electrical wiring harnesses. This happens to be where the maximum bending moment is fundamentally located and essentially reduces the section properties in that region.

It is proposed from the above thought process that the following iterations be analysed using the finite element model:

(a) Enlarge the section size of the frame, particularly at the corner joints, using sizing guidelines for ring beams from Chapter 7, section 7.2 as a starting point. Optimization will be performed later as a minimum size is more desirable for loading cargo.
(b) Plug the holes in the corner joints and assess their effect. Study alternative hole locations away from the high bending moment regions.
(c) Create more overlap of the metal at the joints, essentially stiffening that area.
(d) Reinforce the frame's step discontinuity at the beltline area to assess its effectiveness if the step cannot be smoothed.

If applied in a *synthesis* mode, the process would be in reverse: the free body diagrams from the SSSs mentioned above would have been generated before the start of design and finite element modelling. The initial section size and strategy for maintaining shear flow continuity and joint stiffness would have already been brought to the table and considered among other requirements. The result of this would have likely been a design that, when first analysed, might have fared more favourably and thus incurred less design rework later. Additionally, more effort could have been spent on optimization.

9.5 The context of the preliminary design stage in relation to the overall body design process

9.5.1 Timing

A certain number of weeks are given to develop and rationalize a preliminary design in terms of customer driven and regulatory performance goals, cost constraints, manufacturing philosophy, and weight limitations. Because the inclusion and placement of any structural member has implications to any of the aforementioned items, the preliminary design stage and the decisions which result will set the structural layout of the vehicle. The amount of time allocated will depend on the degree to which the design is new or a modification of an existing design. Time is also influenced by the number of parts and systems involved. The greatest amount of time is for a programme where a family of variants must be developed for a global market, and where 80 per cent or more of the structural content is considered to be new. At the other end of the spectrum is the development of a new variant from an existing platform for a local market where 40 per cent or less of the structural parts may be new. For global programmes, time must be factored in for identifying the customer and regulatory requirements for each market that the common platform has to manage. The requirements and customer expectations may conflict across global marketing regions. Engineering, manufacturing, and cost measurement philosophies may also differ. These create significant issues when the programme is trying to develop and rationalize a vehicle for global markets. The time and energy it takes to work these issues should not be underestimated.

Tools that can help the designer or engineer iterate on many structural proposals quickly have the best chance of impacting the design decisions that need to be made in the preliminary stage. Significant effort can be saved if a collection of various structural configurations is already available that summarize capabilities in meeting performance, weight, cost, and manufacturing constraints. Having actual performance data on hand from previous or current series production vehicles lends credibility. The downside is that it may unduly limit the scope of options that need to be considered, particularly if the weight–cost–performance–manufacturing targets are increasingly challenging and the development cycle time must be shortened. Here an understanding of basic fundamentals, with the help of the SSS method, could help propose and rationalize an alternative structural layout in the early stage. This is also true of benchmarking.

It is common practice in the industry to disassemble and examine competitive vehicles. However, actual performance data to verify the effectiveness of a competitor's structural design may be impossible to obtain without conducting one's own tests of the product, or may be limited to the availability of consumer or government agency data. Even if such data were immediately at hand, it may be misleading to conclude that a particular feature on a competitor's body structure was responsible for a certain total body system performance metric being achieved. This is because a vehicle's total structural performance is dependent in most cases on the interaction among its subsystems, which may not be practical to sort out definitively on a competitor's vehicle without an extensive analysis. Stepping back and rationalizing a competitor's structural features with the aid of the SSS method has the potential to offer fundamentally sound explanations in a relatively short period of time.

For example, idealizing a competitor's structural layout may quickly explain the absence or inclusion of a structural member. If the opportunity exists to measure the vehicle, the SSS method may be used to help explain, from a qualitative standpoint, why the global stiffness is higher or lower than expected.

Other aspects of the preliminary design phase which should not be overlooked are (1) the dominant performance requirements, and (2) the effect of specialization vs integrative (multi-disciplinary) approaches.

1. *Dominant structural criteria.* While all vehicle structure performance criteria have importance, there needs to be priority and focus placed on the 'dominant' or governing load cases which determine the size, shape, and placement of the major structural members. This is because such dominant criteria will influence the vehicle's architecture and therefore must be considered early in the design process. Structural requirements that drive the governing dominant load criteria usually are (a) crashworthiness, (b) durability (extreme load cases), (c) vibration (vehicle body structure's first natural frequency), and (d) vehicle bending and torsional stiffness. Items (b) and (d) are addressed in this book using the SSS method and basic fundamentals.

2. *Specialization vs multi-disciplinary approaches.* The demand to improve performance in response to ever increasing customer expectations and regulatory requirements, while simultaneously reducing cost and weight, has increased the need for computer simulation specialists in crashworthiness, noise and vibration, and durability. The complexity and interactions have made it more challenging to achieve timely and balanced solutions to these issues. Integration functions are often needed to facilitate decision making when solutions conflict. It is therefore

important that the structural engineer be well grounded in the basic fundamentals of the above disciplines in order to be able to synthesize data and contribute to a balanced technical solution. It is suggested that knowledge of the SSS method can provide the body engineer with one of the basic fundamental tools to help sort out and rationalize alternative design proposals.

9.5.2 Typical analytical models (FEM etc.) used at different stages in the design cycle

Only brief mention of this will be made here. Ready availability of commercial hardware and software, and the widespread employment of engineers applying finite element (FE) techniques, are commonplace. It may be logical to describe the analytical modelling process as progressing from simple, qualitative, and coarse in the early stage to fine and detailed as the design progresses. Finely meshed models in the early stage may be hybrid models which incorporate portions of the new design integrated with a previous model where the design is more fully developed. In this manner, direct comparisons between a known quantity and the effect of the new design can be assessed. The early FE 'stick models' that include beam and shear panel elements are somewhat analogous to the idealization behind the SSS method, but can be statically indeterminate. The plate elements carry the loads in the plane of the panel. The beam elements carry the axial forces and bending moments at each end. These models may be used in conjunction with the SSS method to assess and explain the internal load reactions.

Preliminary design and analysis of body subassemblies using the SSS method

10.1 Introductory discussion

In their book *Aircraft Structures* (2nd Edition, McGraw Hill, 1982), Dr David J. Peery and Dr J.J. Azar described the analysis of semi-monocoque structural members using approximate methods. Their discussion of the distribution of concentrated loads to thin webs is applicable to the principles of the SSS method presented in this book for passenger car structures.

Similar to aircraft, modern passenger car structures are constructed primarily of sheet metal. The metal serves as both the structure and primary enclosure. Examples are the floorpan, motor compartment side panel, and rear compartment vertical quarter panels. Idealizations of these structures have been shown in previous chapters.

Other more localized parts of the car structure can be idealized such as supports for chassis and power-train mountings. These support brackets often serve as mounting points for elastomers to isolate noise and vibration. The brackets and surrounding support members must have adequate structural impedance in order for the elastomers to do their job of vibration isolation.

Adapted from thin web solutions described in Peery and Azar's book, the following guidelines should be helpful in executing effective SSSs:

- An SSS can resist only tension, compression, and shear forces *in its own plane*.
- Stiffeners (either integral to the SSS or add-on) are required to resist *compression* forces in the plane of the SSS.
- Similarly, stiffeners are required in order to resist *small*, distributed loads *normal* to the SSS.
- *Large* concentrated loads must be resisted by transmitting loads to the plane of an *adjacent* SSS.
- An SSS is necessary for each significant perpendicular component of the load, such that the loads are applied at intersections of two perpendicular SSSs – or else

additional members must be provided to span between and transfer loads to the intersecting SSSs.

- When a large concentrated load is applied in the plane of the SSS, a stiffening member should be incorporated to distribute the load.

Some of the above principles can be illustrated with a rear end structure that incorporates a luggage floor (rear compartment panel), wheelhouse inner/quarter sidewall panel, longitudinal beam members, and cross-member. See Figure 10.1.

The luggage floor longitudinal member is used to illustrate the above points by example. This member is the primary load carrying structure for the rear suspension. The longitudinal beams and suspension cross-members act as stiffeners for the rear compartment panel that is considered to be a SSS. The luggage loads are typically distributed and may amount to as much as 100 kg. This is small when compared to, say, the suspension loads that can amount to a load factor of $3\,g$ (or three times the vehicle weight that is distributed to a particular wheel corner). Severe pot-holes and other extreme road disturbances can easily double this load factor.

Looking at a cross-section of the longitudinal rail structure in Figure 10.2, the rail hat section is closed out by the rear compartment SSS panel. The section shown is where the rear suspension jounce (bump) loads are reacted. The edge corners are considered stiff. However, because the load is acting perpendicular to the bottom flange of the beam, the surface easily deflects. Excessive deformation would result in degradation of the available suspension ride travel and compromise the structure's ability to sustain further loading.

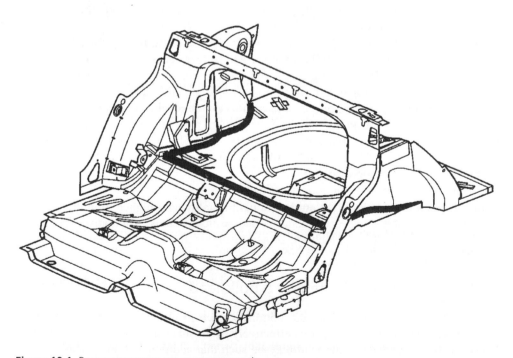

Figure 10.1 Rear compartment pan structure example.

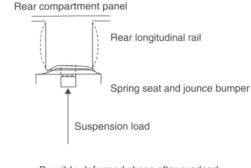

Figure 10.2 Cross-section of rear longitudinal rail at spring seat.

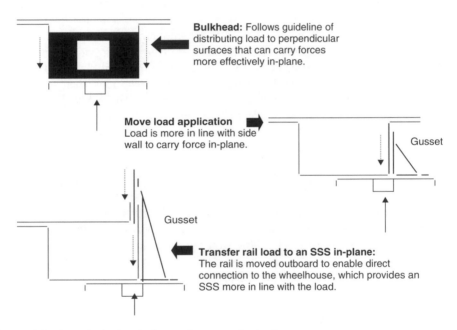

Figure 10.3 Alternative load paths for reacting suspension load.

Just thickening the part might not be enough. Employing a higher strength material may not be allowable under manufacturing and cost constraints. Alternative load paths are suggested, examples of which are shown in Figure 10.3.

10.1.1   Alternative 1: employ a bulkhead

This solution transfers load from the bottom surface to the perpendicular web sidewalls, which can carry in-plane loads effectively. Both sidewalls carry the load evenly. However, the magnitude of the load may be such that additional countermeasures are needed.

10.1.2   Alternative 2: move where the load is applied to a more favourable location

This solution puts the application of the load closer to the stiffer web sidewall. The disadvantage is that twisting may be induced in the rail which must be resisted by an adjacent member, such as a crossbar. Even so, this may not be sufficient as the web sidewall may be overstressed. Again, more countermeasures may be needed to manage high overload conditions such as severe pot-holes and bumps. A gusset bracket may need to be added in order to buttress and stabilize the sidewall.

10.1.3   Alternative 3: transfer the load to an SSS perpendicular to the rear compartment pan

Note that this solution introduces the wheelhouse inner as a major load path by providing direct connectivity to the beam. The beam no longer has to depend on the relatively soft rear compartment pan for support. This will provide a satisfactory solution if the path from the wheelhouse inner to the 'C'-pillar inner is stiff. If it is not feasible to move the beam outboard because of other considerations, then an auxiliary bracket may be provided to transfer load into the wheelhouse inner SSS.

A vertically oriented stiffener will need to be provided to the wheelhouse inner panel in the form of a swage (bead) or add-on reinforcement. Care must be taken to avoid stress concentrations that might lead to fatigue cracks.

The above examples were meant to illustrate some of the guiding principles stated in the introduction. Applying the guidelines will enable fast qualitative design direction and visualization of alternatives. Once the alternatives are filtered down during concept selection, detailed finite element analyses can be employed to optimize and confirm the design. The preceding exercise provides a basis from which the later finite element analysis results can be assessed – i.e. how well are the loads getting transferred from the primary members to the adjacent structure? The importance of drawing free body diagrams *first* in order to get a feel for the problem cannot be overstated. A procedure may be listed as follows:

1. Draw the free body diagrams (FBD) which idealize the overall problem as simply as possible.
2. Idealize/visualize the problem as SSSs.
3. Utilize guidelines provided in section 10.1 to assess and provide design concept alternatives.

10.2   Design example 2: steering column mounting/dash assembly

10.2.1   Design requirements and conflicts

The steering column mounting structure is considered to be an integral part of the front-of-dash (FOD) or driver's 'cockpit' assembly. An example is shown in Figure 10.4.

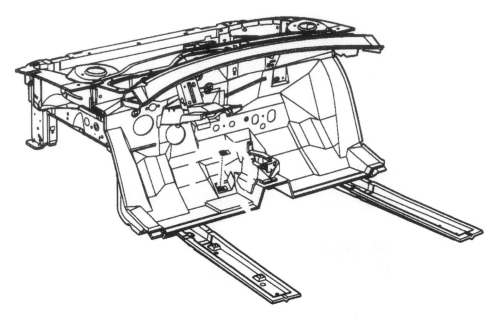

Figure 10.4 Steering column support and dash assembly.

The steering column support structure is usually attached to the dash and cowl panel. It may be augmented by attachment to a transverse beam that is either cantilevered off or supported between the front body hinge pillars (FBHPs). In addition to supporting the steering column, the structure may also have to support the brake pedal pivots, clutch pedal pivots and instrument panel. The primary structural performance criteria for the steering column assembly are usually:

1. Meet a minimum natural frequency target to assure vibration isolation from road and engine idle excitation.
2. Accommodate occupant safety and vehicle crashworthiness objectives. Crash requirements are beyond the scope of this book, so only the stiffness criteria will be considered here.

10.2.2   Attached components

Figure 10.5 depicts a typical side view arrangement for the steering column support assembly.

In this example, the steering column is considered to be a beam of more or less uniform circular cross-section. The lower forward end of the column typically allows for pivot adjustment in side view and is considered to be simply supported. This end also connects to an intermediate shaft which rotates and connects to the steering rack. The upper rearward end of the column is usually bolted to a support bracket and may be considered either clamped or simply supported. The steering wheel is cantilevered rearward from this position. Its location is determined by the occupant package criteria. In simple terms, it is considered to be weight hanging from a supported beam.

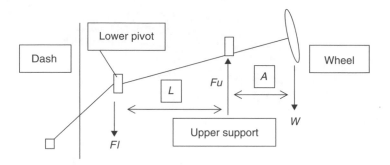

Figure 10.5 Side-view diagram of steering column mounting scheme.

A free body diagram is drawn to visualize the reaction forces at the upper and lower supports. For simplicity, the side view tilt angle is assumed to be small such that no cosines need to be determined. The lever arm of the wheel's mass is reacted by a counterbalancing moment generated by the forces at the upper and lower pivot supports. The equation for vertical deflection at the wheel is:

$$Y = -WA^2(L + A)/3EI$$

from *Mechanical Engineering Design* by Shigley (1972), where Y = vertical deflection of the steering column beam at the steering wheel, W = weight of steering wheel, A = cantilever distance of wheel from upper steering column support, L = distance between upper and lower supports, E = elastic modulus and I = second moment of area of the steering column.

Engine idle excitation and road bump inputs generate forces through the body structure load paths which cause steering wheel motion at the end of the column beam. Excessive motion or shake is perceived as poor ride quality by the driver. It is therefore desirable to minimize Y, and thus optimize the beam's resistance to steering wheel shake. The above equation indicates that the following will directionally result in reducing Y:

1. Reduce steering wheel weight W. Constraints on wheel size, use of alternative materials and airbag requirements are among factors to consider.
2. Reduce A, the cantilever distance from the upper support point to the wheel. *This is perhaps the most significant dimensional factor* because it is a squared term. Intuitively, one can visualize it as analogous to a simple cantilever beam problem. The occupant seating arrangement and component packaging requirements are among the constraints.
3. Reduce $L + A$, the total distance between the wheel and the furthest support. Intuitively, it would be desirable for L to be greater than A. The same constraints as (2) generally apply.
4. Increase the steering column beam bending stiffness EI. Assuming a given material selection, this would require increasing the thickness and/or diameter. Mass, cost, and packaging constraints need to be considered.

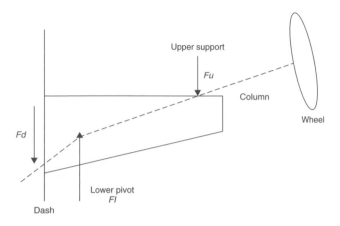

Figure 10.6 Reaction forces on the steering column support bracket.

Once the above factors have been optimized, the reaction forces and support bracket structure need to be addressed. The forces on the adjacent support bracket structure shown in Figure 10.6 will be equal and opposite to those shown in Figure 10.5.

There will be two symmetric support brackets, one on each side of the steering column axis. Note that the dash panel is a reactive surface. Although the forces are carried in two SSSs, it is intuitively obvious that two light gauge sheet metal brackets cantilevered from the dash panel alone do not provide a robust design solution. Before alternatives are considered, the fundamental forces are determined.

From Figure 10.6:

Summing forces: $-W-Fl+Fu = 0$
Summing moments about the lower pivot: $-W(L+A)+FuL = 0$

$$Fu = W(L+A)/L$$

Note that the distance from the upper steering column attachment to the steering wheel mass is the cantilever dimension A. A shorter cantilever will result in less reaction force that must be managed.
Substituting:

$$-W-Fl+W(L+A)/L = 0$$

$$Fl = W(L+A)/L - W$$

From Figure 10.6:

Summing forces: $-Fu+Fl-Fd = 0$
Substituting:

$$Fd = W(L+A)/L - W - W(L+A)/L = -W$$

From the above equations it is apparent that (a) the lower pivot reaction force Fl is small compared to the upper reaction force Fu, and that (b) the vertical dash plane reaction force Fd is equal in magnitude to the steering wheel weight W. As far as the support structure is concerned, the problem can be simplified by setting Fu equal to W, or $Fu = W = -Fd$. This will allow the problem to be treated similarly to the

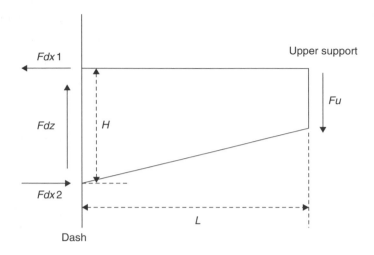

Figure 10.7 Simplified loading for steering column support bracket.

manner in which suspension loads are resolved into the body from previous chapters. Figure 10.7 depicts the more simplified loading condition.

Summing forces:

$-Fu + Fdz = 0$. This creates an unbalanced moment. Horizontal forces must be introduced.

Summing moments about the lower forward connection at the dash:

$$-FuL + Fdx1H = 0$$

$$Fdx1 = FuL/H$$

And by equilibrium:

$$Fdx1 = Fdx2$$

It is apparent that:

(a) The horizontal forces $Fx1$ and $Fx2$ need an SSS to react against. The dash panel is perpendicular to the forces and by definition will not function as an SSS. It will act as a flexible membrane. Beading, swaging, or a local reinforcement plate will increase its stiffness, but they are inherently inefficient substitutes for an SSS. Some local stiffening might also be afforded by the plate that locally reinforces the dash while functioning as the attachment area for the brake booster assembly.

(b) In order to minimize the loading perpendicular to the dash panel, it is desired to maximize the vertical span of the bracket (H) and minimize the cantilevered length of the bracket (L). The vehicle's cowl height, the required location of the pedals, occupant seating position criteria and component packaging criteria are the constraints that will need to be considered.

Alternative A: provide an SSS by utilizing the cowl air plenum panel

Figure 10.8 illustrates the utilization of the cowl air plenum panel as an SSS to react the horizontal loads.

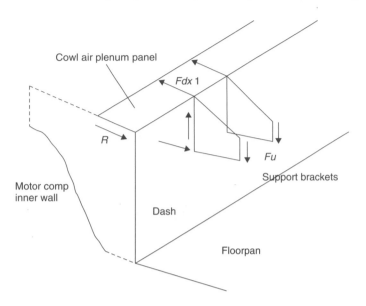

Figure 10.8 Providing an SSS to react horizontal forces at the dash: alternative A.

Such an arrangement is feasible when the support bracket can be packaged very close to the corner where the plenum and dash panels meet. The reactions are at the ends of the panel where it joins the motor compartment inner shear wall. This may provide a good solution for reacting the upper force $Fdx1$, but $Fdx2$ will be more of a challenge because there is no nearby SSS to utilize. This area may have to be carefully stiffened locally with a reinforcement, swaging and/or beading.

Alternative B: provide an SSS by utilizing the cowl bar beam

Figure 10.9 illustrates the utilization of the cowl bar beam as an SSS to react the horizontal loads.

The cowl bar is usually a double hat cross-section, curved beam running cross-car between the front body hinge pillars (FBHPs). It also carries shear loads along the windshield bottom periphery during body torsion. The bottom half of the beam could be regarded as an SSS. It has the benefit of utilizing the stiffness of a closed box cross-section. By contrast the cowl air plenum panel is usually characterized as an open section because its upper part has considerably sized holes cut out for air flow. The reactions to the cowl bar SSS loads are provided by the connection to the FBHP inner shear wall. Body fabrication and vehicle assembly process constraints will usually dictate how the steering column support bracket can be attached to the cowl bar. Steering column vertical loads will also induce twist in the cowl bar. Therefore, a large enclosed sectional area, minimum cut-outs and discontinuities, and stiff end connections are desirable.

Alternative C: provide an alternative load path with an auxiliary beam

Figure 10.10 shows the addition of a transverse beam attached to the rearward end of the support brackets.

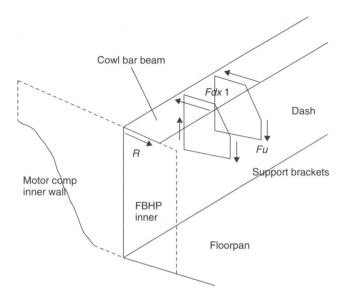

Figure 10.9 Providing an SSS to react horizontal forces at the dash: alternative B.

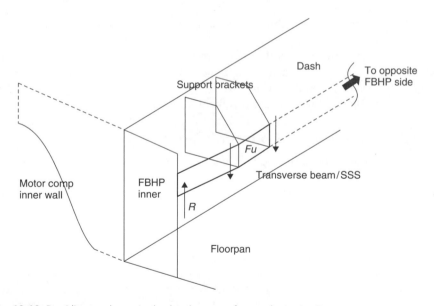

Figure 10.10 Providing an alternative load path to react forces: alternative C.

This beam can be idealized as an SSS. It sees loading from Fu, the upper support bracket forces. The reaction is provided at the FBHP inner wall SSS. This arrangement will produce the equivalent of a simply supported or cantilevered beam. The beam/SSS can be either full width cross-car between FBHPs or cantilevered and supported from the driver's side FBHP. This member also provides a means to support the instrument panel (IP). Another benefit is that this member acts to tie the two support brackets together. Note that if the centroid of the beam is not in line with the steering column upper attachment, torsion will result (Figure 10.11), which makes it necessary to have a box section of sufficient enclosed area. In addition a stiff connection to the FBHP is recommended.

This may be accomplished by either (in Figure 10.12) (1) a bolt-through attachment which utilizes both the inner and outer FBHPs as SSSs or (2) bolting to the web of the FBHP near where it intersects the flange.

If it is not desired to have the beam react torsion, then the scheme shown in Figure 10.13 may be employed.

This may be the case when there is insufficient space to provide a member with adequate enclosed area to resist twisting. In effect, the beam is carrying the upper attachment load instead of the support bracket. The bracket is carrying the lower load.

In either case, the beam section can be initially sized according to a stiffness target or a minimum natural frequency requirement. It will need to be determined if the section size is governed by stiffness as mentioned above, or by ultimate strength (i.e. for crashworthiness).

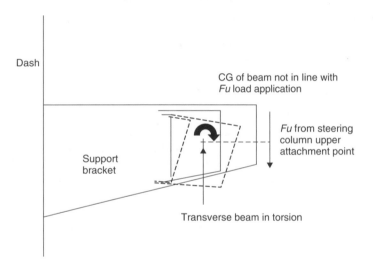

Figure 10.11 Side view of auxiliary transverse beam.

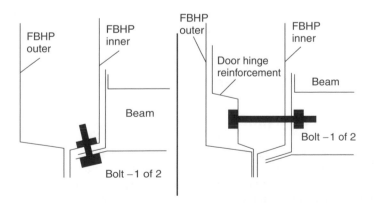

Figure 10.12 Considered methods of attaching transverse beam to FBHP (plan or overhead view).

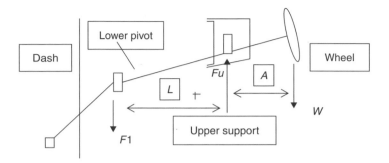

Figure 10.13 Cross-member used as beam without torsion.

10.3   Design example 3: engine mounting bracket

The engine or power-train mounting scheme provides several significant functions:

1. Supports the engine and transmission weight.
2. Provides reaction to drive-train torque.
3. Controls power-train motion.
4. Provides isolation from noise and vibration.

A discussion on any of the above functions would be lengthy and beyond the scope of this book. The purpose of this section is to illustrate the application of SSS principles to the design of support brackets. In this example a transverse front wheel drive power-train is assumed. Examples of a typical mounting scheme are shown in Figure 10.14. An idealization is shown in Figure 10.15. Typical loadings are shown in Figure 10.16. It should be noted that the load factors given are meant to be illustrative examples, not absolutes. Each company will utilize their own criteria based on experience with particular power-trains and mounting schemes. In any event, load factors used for preliminary design should be validated with test and/or simulation data as more information becomes available.

Loads are transmitted from the engine to the bushing which is attached to the body mount bracket. The bushing is composed of a housing which encases an elastomer. The main point is that the bracket must react loads in all three translational directions. In addition to the loads shown, the bushing is responsible for providing noise and vibration isolation. The stiffness of the bracket and surrounding structure must be sufficiently high for the bushing to provide effective isolation. Each translational load direction will be addressed separately.

10.3.1   Vertical direction

This is the primary load bearing reaction. Approximately half the weight of the engine/transmission assembly is supported by this mount. It may have to sustain load factors as high as three or more during road bumps and pot-hole disturbances. In the case of 4-cylinder inline power-trains, there must be sufficient isolation from the inherent vertical shaking forces. An examination of the loading diagram in Figure 10.17

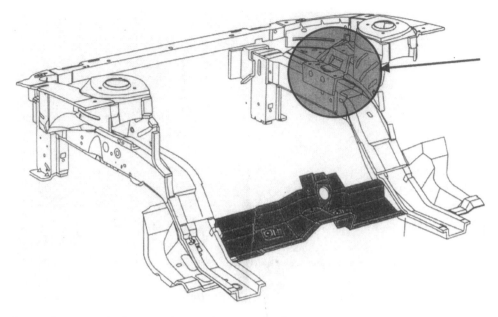

Figure 10.14 Example of power-train mounting support brackets.

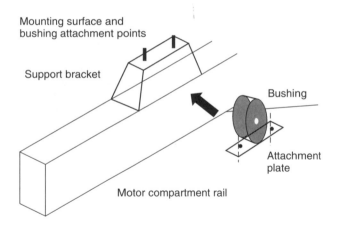

Figure 10.15 Engine mounting-to-body example.

shows there is an opportunity to connect the bracket to the *inboard* side of the motor compartment rail.

This will induce shear flow in the motor compartment rail and provide an in-plane reaction. Figure 10.18 depicts one possible design alternative.

The bracket is also given the shape of an arch for structural efficiency. Close examination reveals that the *outboard* side of the bracket cannot be connected to the motor compartment rail outer sidewall. The path is blocked by the motor compartment inner panel. Cutting holes in this panel to provide weld access would defeat its purpose of providing an SSS reaction to the front suspension loads and would also create stress

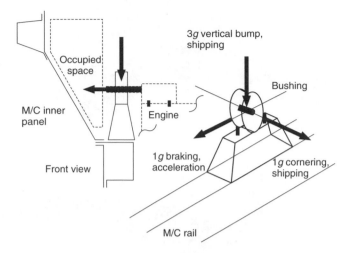

Figure 10.16 Typical loads on engine mount bushing.

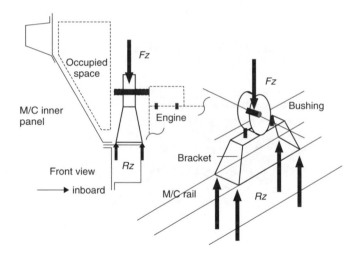

Figure 10.17 Simplified engine mount vertical load diagram.

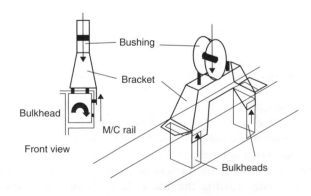

Figure 10.18 Potential design alternative for bracket vertical direction load.

risers. Adding a member inside the dotted line zone is prohibited in this example because of assumed interference with other functional devices such as coolant recovery bottles, lines, and hoses. Attaching to the top surface of the motor compartment rail alone would not achieve an SSS because the loading is normal to the rail surface. However, incorporating a bulkhead inside the rail would facilitate in-plane load transfer to the rail inboard side wall and thus achieve an SSS. The bulkhead cannot be attached to the outer sidewall because of the assembly sequence. The bulkhead is welded to the top and inboard side wall surfaces with flanges. The engine mount bracket loads the bulkhead through a flange positioned directly in line with the bulkhead with three-metal thickness spot welds. Each of the two bulkheads shown in Figure 10.18 also welds to the bracket's sidewall extension. Note that the reaction force produces an unbalanced moment that will need to be resisted by the torsional stiffness of the rail section. Therefore, sufficient enclosed area from the rail section will be necessary.

10.3.2   Lateral direction

Besides reacting cornering loads and motions during shipping and transport, this direction can also be important for noise and vibration isolation. As far as the body mount bracket is concerned it will be the most difficult direction to manage because it becomes in effect a cantilevered beam. The occupied space outboard of the bushing precludes adding a structural member for support. The first consideration should be whether the cantilever can be made shorter. Another consideration should be whether the load can be transferred to another mount. On a transversely mounted power-train, this might be accomplished by making the engine mount elastomer relatively soft and stiffening the transmission mount. The trade-offs might be reduced isolation and increased stress in the transmission mount.

The body mount bracket may be idealized as two cantilevered beams sharing the load. The required moment of inertia property may be calculated from a stiffness that is derived from a structural impedance specification. It is found that the legs of the bracket will need to be closed out or boxed in order to achieve the required inertia property. This is provided by a close-out part that also connects to the rail. The problem then remains to assess how adequately the moment can be supported at the 'fixed' end of the beam.

Figure 10.19 shows that the lateral load reaction at the bottom of the bracket will induce shear flow and torsion in the rail.

As with the vertical load case, the rail will need sufficient enclosed area to resist twisting. Each bulkhead will tend to act as a stabilizer but fabrication processes in this hypothetical case restrict attachment to no more than two rail surfaces.

10.3.3   Fore–aft direction

This situation is somewhat analogous to the body sideframe loading idealization. The fore–aft load will be shared by the two inclined beams or pillars which are the legs of the bracket. The bracket may be thought of as an 'A' frame. Stiffness and stress for each leg can be verified using the worst case assumption of two cantilevered beams. These legs are assumed to be previously sized from the lateral load case. The

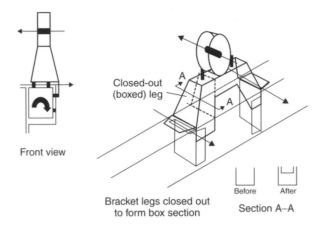

Figure 10.19 Design alternative modified for lateral direction load.

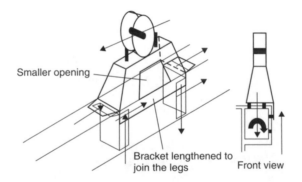

Figure 10.20 Design alternative modified for force–aft direction load.

applied fore–aft load produces an equal and opposite reaction at the rail. The resulting moment is balanced by two equal and opposite vertical reactions, one at each end of the bracket. The rail's top flange and inboard sidewall web provide good in-plane surfaces to support the fore–aft load. The bracket's overlapping surface is lengthened and the 'A' frame's centre opening made smaller in order to provide a more stable shape, as shown in Figure 10.20.

The vertical reaction is best supported by the rail's inboard sidewall, which is stiffened by the bulkhead and the bracket's overlapping surface. This force will induce shear flow in the rail. Torsion will be induced in the rail as a consequence of the vertical force, as evidenced by the front view shown in Figure 10.20.

10.3.4 Summary

The above design alternatives are only one subset of many possibilities. It is also apparent that the realities of packaging and manufacturing constraints challenge the ability to achieve the 'ideal' structure. In the engine mount case, relatively good SSS in-plane reactions to external loads were achievable on the inboard side, but not the outboard. For these situations, it became important for the rail to be stiff in torsion, with sufficient enclosed sectional area and close weld spacing along its length. At the beginning of the problem, the only givens were the location and direction of the applied loads, with some packaging constraints. At the end of the problem, a qualitative design concept with shapes was produced from which more detailed development could begin.

10.3.5   Discussion

It has been shown how SSS analogies and basic engineering fundamentals can be combined to (a) visualize load paths, (b) do preliminary member sizing and (c) generate concept alternatives for consideration. In the early stage, finite element models of simple structural surfaces may be constructed within the package space envelope to provide quick and early comparisons of various concept proposals. Once a viable concept is selected and design data are generated, finite element analyses can be used to optimize and verify the design before hardware is committed. SSS analogies and basic engineering fundamentals are again employed to help interpret and understand the results. Even if the SSS analogies and basic engineering principles are used to provide only *qualitative* direction, the time invested should pay dividends downstream in the form of CAD and CAE resources being more effectively utilized to develop fundamentally sound concepts. Had the situation been where an existing design was the issue and remedial action needed, the same principles could be applied. First, the existing structure would be idealized, followed by free body diagrams to visualize how the loads were being carried. This fundamental understanding is a prerequisite to more efficient and effective problem solving.

10.4   Design example 4: front suspension mounting

The detailed loading on the local structure from an independent suspension requires careful study because there are several load cases to consider and the type of suspension system determines the attachments to the structure. As an example, the double transverse link suspension (double wishbone) with a subframe will be studied.

10.4.1   Forces applied to and through the suspension

Three local cases will be considered. First, the kerb bump case where a high lateral load is applied at the tyre/ground contact area and is combined with the vertical load. For this example, the lateral load is taken as twice the vertical load on the wheel. (Individual vehicle designers may take different load factors.) Figure 10.21(a) illustrates this condition with a resultant load of $2.236\,W$ where W is the vertical load. Considering the wheel/stub axle as a body, it is held in equilibrium by forces at C and B, the top and bottom pivot points. Using the principle that three forces acting on a body in equilibrium are concurrent, the point of concurrency is O_1 which must be on the line CD produced as the top link CD cannot have any lateral force because it is simply a strut or tie bar. The lower link AGB has lateral forces applied by the spring/damper and by components of the forces at A and B. Therefore the force at C is along the line CO_1 and the force at B acts along BO_1.

The force vector diagram for the wheel/stub axle is shown in Figure 10.21(a) which enables the force F_{Bk}, the force at B due to curb bumping, and the force in the top link F_{CDk} to be evaluated.

Applying the same principle for the equilibrium of the link AB, the point of concurrency is O_2 and the force vector diagram can be drawn (again see Figure 10.21(a)). These drawings, if drawn accurately, enable the values to be obtained. Note that F_{Bk} and F_{Ak} are shown to be about twice the wheel vertical load.

Figure 10.21(b) shows the forces applied to the wheel/stub axle, the two links and to the vehicle body or subframe. The force at A is largely horizontal and about 2.5 times the static wheel load, the force at F is similar to the wheel load and at D the force is about half the wheel load.

The second load case is the very high vertical bump load which is often taken as three times the static vertical load. Using the principle of three concurrent forces again the points of concurrency are again O_1 and O_2 as illustrated in Figure 10.22(a). Again,

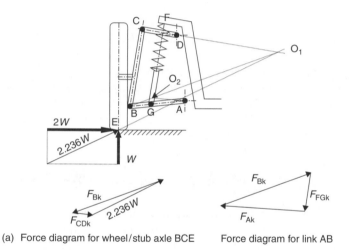

(a) Force diagram for wheel/stub axle BCE   Force diagram for link AB

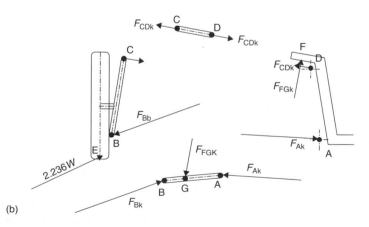

(b)

Figure 10.21 Double transverse link suspension (kerb bump).

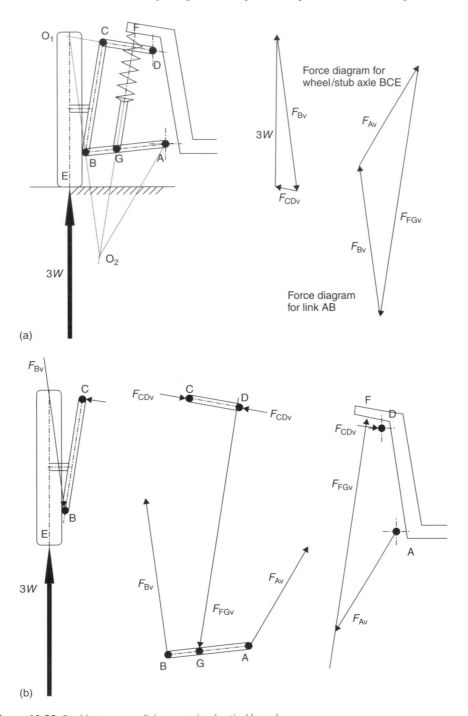

(a)

(b)

Figure 10.22 Double transverse link suspension (vertical bump).

the force vector diagrams are shown indicating a small force in the top link F_{CDv}, large forces FA_v and F_{Bv} at the ends of the lower link and a very large force in the spring/damper F_{FGv}.

Figure 10.22(b) shows the forces applied to the wheel/subaxle, the two links and to the vehicle body or subframe. In this case the force at the spring/damper mounting F is very large of the order of five times the static wheel load. The force at A is applied downward at an angle and is approximately 2.5 times the static wheel load while the force at D is similar in magnitude to the first case but in the opposite direction.

A third load case is the braking case which can be considered by the approximate method shown in Figure 10.23. By taking moments about B the force:

$$F_{Cb} = W(r - a)/(a + b)$$

and by resolving forces:

$$F_{Bb} = W + F_{Cb}$$

where W is the braking force which can be taken as equal to the dynamic vertical wheel load including weight transfer due to braking.

Then as shown in Figure 10.23 considering the forces on each suspension link the lateral forces acting on pivots A and B can be found by taking moments:

$$F_{Ab1} = F_{Ab2} = cF_{Bb}/d$$

$$F_{Db1} = F_{Db2} = eF_{Cb}/f$$

The longitudinal forces F_{Bb} and F_{Cb} can be assumed to be shared by the front and rear bushes of the pivot axis.

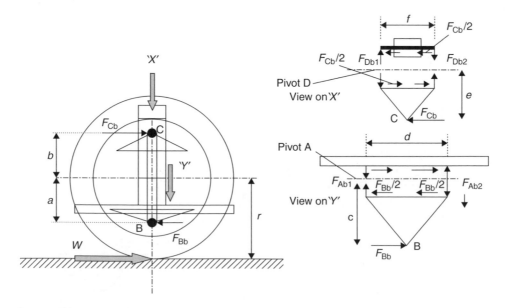

Figure 10.23 Double transverse link suspension (braking).

10.4.2   Forces on the body or subframe

The forces developed by these load conditions on a subframe are summarized in Figure 10.24. while Figure 10.25. shows an illustration of an actual suspension and subframe. This type was quite common in vehicle design a few years ago when the length between the top and bottom joints of the stub axle/knuckle was small. The whole suspension and the main engine mounts were then incorporated in a relatively small subframe that was attached to the body at four points, one at each end of the longitudinal members.

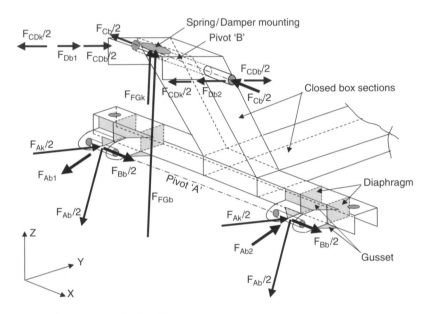

Figure 10.24 Front suspension loads subframe.

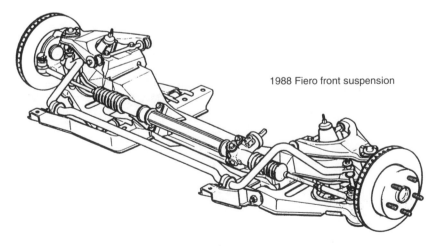

Figure 10.25 Double transverse link suspension and subframe (from Automotive Engineering, October 1987).

Applying the SSS method to the longitudinals, the lateral components of load will require an SSS in the horizontal $X-Y$ plane while the vertical components will demand an SSS in the vertical $X-Z$ plane. The requirements of these two SSSs are incorporated in the one beam shown as a channel section. The vertical component of the forces $F_{Ab}/2$ are, however, offset from the vertical beam and will cause a torsion moment in the beam. Open section channel sections are flexible in torsion so there is the need to add diaphragms adjacent to the pivot brackets and possibly to make the beam into a box section. The pivot brackets will need gussets due to the longitudinal forces $F_{Bb}/2$ as these are loaded normal to their plane which again is against the rules of SSSs.

The large force F_{FGb} from the spring/damper will cause the inclined member to bend and be in tension. Therefore an SSS is required in the $Y-Z$ plane. The longitudinal forces $F_{cb}/2$ will cause bending approximately in the $X-Z$ plane while the lateral forces will cause torsion. A closed box section is therefore required for this inclined member in order to carry all the different forces and moments. This closed member is then extended across the vehicle to carry engine loads and the steering rack. Stiffness of a box section is essential in this area to ensure that steering geometry is precise and to provide satisfactory engine supports.

Later designs of double transverse link suspensions have a longer spacing between the top and bottom joints of the stub axle/knuckle. This is often decided by the inclusion of

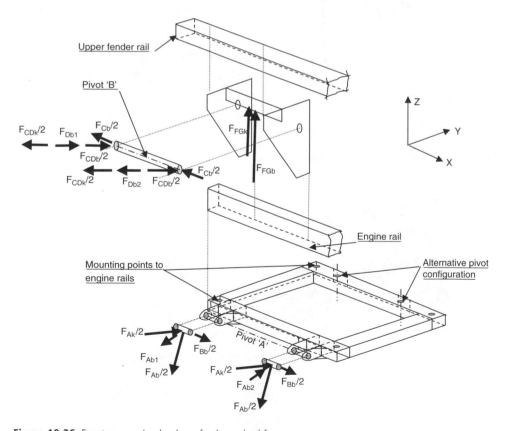

Figure 10.26 Front suspension loads on fender and subframe.

a drive shaft requiring a high position for the spring damper or by suspension geometry characteristics. The high mounting for the spring results in the spring loads and the top link loads being applied directly to the structure rather than to a subframe. Figure 10.26 shows an SSS model of the suspension tower that is mounted between the upper fender rail and the engine rail or longitudinal at the lower inner panel of the fender.

As Figure 10.26 shows, there are loads applied to the suspension tower in all three orthogonal directions. This requires the upper fender rail and engine rail to carry bending moments in the $X-Y$ and $X-Z$ planes, shear in both planes and axial loads. This results in the need for closed box members as indicated.

The subframe now is a simple rectangular peripheral frame mounted to the main structure at the four corners. If the lower suspension arm is mounted as shown in Figure 10.26 the structural requirements for the side members remain similar to the previous design except the bending in the vertical $X-Z$ plane is reduced as the spring loads have no influence on this subframe. Due to the pivot 'A' axis being outboard of the side members, there will also be torsion on the side members and in turn bending on the front and rear cross-members. Therefore, the subframe is made from closed sections for all four components. A practical subframe and suspension system is shown in Figure 10.27. The detail design is governed by structural, chassis and local package requirements.

A further modern development is to replace the pivot A bushes with bushes mounted with their axis vertical (along the Z-axis). This is done to reduce the risk of the

Figure 10.27 Mercedes-Benz S-class double transverse link suspension and subframe (from *Automotive Engineering*, November 1998).

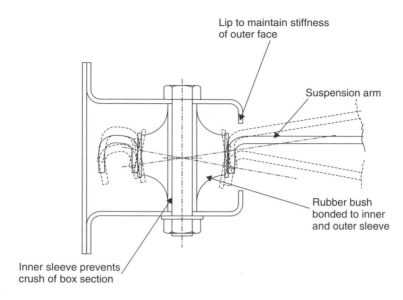

Figure 10.28 Local section through suspension bush and subframe.

suspension bushes rotating in their housings and generating an annoying squeak. In this case the outer face of the side members must have openings to permit the arms of the lower suspension link to pass into the side members. The side members are then not subject to such a large torsion moment but the openings in them can reduce their bending stiffness. It is important to maintain the stiffness of the outer face by ensuring the aperture leaves a lip so that it maintains the properties of an SSS. Figure 10.28 illustrates the principle of this system of bushes and a cross-section through the subframe side member. The inner sleeve of the rubber bush is necessary to prevent normal loads due to bolt tightening on the top and bottom surfaces (SSSs) which will crush the box section. The mounting of the subframe to the vehicle structure must similarly ensure no crushing of the box section, one design alternative is to use spacer tubes.

From this example of a suspension mounting it is possible to see how the principles of the SSS method can assist in designing the details of the local structure around the suspension to structure attachments. Further load cases, for example, vertical bump loads combined with lateral cornering forces, may need to be considered. The cases described in this section are sufficient to show that the force vector diagrams combined with the SSS method can provide useful information on the magnitude of loads and features that must be incorporated in the local structure.

11

Fundamentals and preliminary sizing of sections and joints

Objectives

- Overview of section and joint structural behaviour in a design context.
- To outline the design and sizing approach for sections and joints.

11.1 Member/joint loads from SSS analysis

In Chapters 5, 6 and 7 it was described how to determine the loads on various SSSs and the load path between components. This chapter now considers the sections and section properties that are necessary to carry the loads that have been determined.

11.2 Characteristics of thin walled sections

Thin walled sections can be classified into two types: open section where the thin sheet metal is formed with a discontinuity, and closed section where the section forms a complete loop. Both types of section can be manufactured with more than one component and joined together by spot welds, seam welds or by an adhesive.

11.2.1 Open sections

Typical sections of this type are shown in Figure 11.1. At (a) the angle section is not a suitable structural member when considered alone. The principal axes $U-U$ and $V-V$ are inclined to the faces of the angle and if bending is applied about either $Y-Y$ or $Z-Z$ bending will occur about both axes. Also the stress distribution is very asymmetric and results in large parts of the section being understressed. This is inefficient use of material.

At (b) the Z section has similar undesirable characteristics and hence is not suitable as a section on its own. The principal axes of the Z section are inclined so that bending

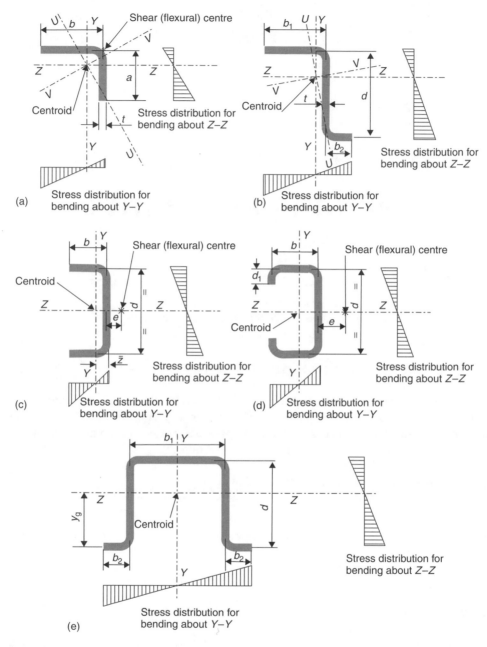

Figure 11.1 Open sections—properties.

will occur about both the $Z-Z$ and $Y-Y$ axes when moments are applied in either plane. The angle of the principal axes $U-U$ and $V-V$ relative to the $Y-Y$ axis will depend on the relative lengths of b_1, b_2 and d.

The channel section shown at (c), however, is a suitable structural section and is used especially in commercial vehicle chassis and body structures. This section is very suitable for bending loads causing moments about the $Z-Z$ axis. It is less satisfactory for bending about the $Y-Y$ axis because it is less stiff (has lower value for I_{yy} than I_{zz}) and has an asymmetric stress distribution. Care must be taken to ensure that the flange width b is not excessive as this can lead to reduced allowable compressive stress. The wide flange results in a low stress at which buckling occurs. Improvement in buckling stress can be achieved by adding a lip to the channel as shown at (d).

The hat section shown at (e) has good bending properties about both $Y-Y$ and $Z-Z$ axes provided the value of $2b_2$ is approximately equal to b_1. As shown, b_1 is larger than $2b_2$ and hence the $Z-Z$ axis moves nearer the top and results in lower stress in the top of the hat when bending occurs in the $Y-Y$ plane. An important feature of open sections is the effect of the shear centre. The channel section shown at (c) has the shear centre as indicated. Den Hartog (1949), Gere and Timoshenko (1991) and Ryder (1969) deal comprehensively about the evaluation of the offset. Here it is sufficient to say that in order to prevent twisting of the section when loaded in shear and bending the shear force and bending moment must be applied through the shear centre.

The offset of the shear centre from the centre of the web (Ryder 1969):

$$e = \frac{b^2 d^2 t}{I_{zz}}$$

where I_{zz} is the second moment of area of the section about $Z-Z$ and b, d and t are defined in the figure.

The major limitation of all open sections is their lack of torsional stiffness due to their very low polar second moment of area. Engineer's torsion theory (Ryder 1969) is not satisfactory because under torsion transverse sections do not remain plane, there is axial displacement at various points on the section. This is known as wharping, and detailed analysis of this is beyond the scope of this book. (Megson 1999) gives a comprehensive treatment of this subject.

The polar second moment of area J_x, for the sections shown in Figure 11.1 are:

$$\text{Angle (a) } J_x = (a + b)\frac{t^3}{3}$$

$$\text{Z section (b) } J_x = (b_1 + b_2 + d)\frac{t^3}{3}$$

$$\text{Channel (c) } J_x = (2b + d)\frac{t^3}{3}$$

$$\text{Channel (d) } J_x = (2d_1 + d + 2b)\frac{t^3}{3}$$

$$\text{Hat (e) } J_x = (b_1 + 2b_2 + 2d)\frac{t^3}{3}$$

In all these examples the thickness of the material t is small and hence the term $t^3/3$ is also small leading to a low value of J_x. This in turn leads to a large angle of twist

θ along a length L:

$$\theta = \frac{TL}{GJ_x}$$

where G is the shear modulus and T is the torsion moment.

The bending stiffness of the open sections shown in Figure 11.1(c), (d) and (e) is a function of their second moments of area. If we take the simple channel section of Figure 11.1(c) the second moment of area about the $Z-Z$ axis is:

$$I_{zz} = \frac{td^3}{12} + 2\left(\frac{bt^3}{12} + \frac{btd^2}{4}\right)$$

Note this ignores the small effect of the bend radial. Numerically this results in a reasonably high value for I_{zz} due to terms in d^3 and d^2.

The value for the second moment of area about the $Y-Y$ axis will have a moderate value given by:

$$I_{yy} = 2\frac{tb^3}{3} + \frac{dt^3}{3} - (2b+d)\,tz^2$$

Again there is a term in b^3 that has a significant influence on the magnitude of the second moment of area. Similar significant values for I_{zz} and I_{yy} are obtained for the sections shown in Figure 11.1(d) and (e). Therefore channel and hat sections can have acceptable bending properties.

11.2.2   Closed sections

The open sections described in the previous section can be combined together or with flat plates etc. to form closed sections. Examples of closed sections are given in Figure 11.2. At (a) there are two Z sections with unequal length flanges joined to form a closed rectangular section. This like the individual Z sections has principal axes that are inclined to the faces of the plates due to the asymmetry of the section. The second moments of area about the $Y-Y$ and $Z-Z$ axes are much increased over the open section.

At Figure 11.2(b) there is the combination of two channels, one with wide and one with narrow flanges. This combination avoids inclined principal axes and still has substantial second moments of area about $Y-Y$ and $Z-Z$ axes.

At Figure 11.2(c) there is the combination of two channels with narrow flanges and two flat plates. This has similar structural properties to the double channel shown at (b).

In Figure 11.2(d) there is a hat section with a flat closing plate while at (e) two hat sections are combined, both of these form effective structural members with good bending properties about $Y-Y$ and $Z-Z$ axes.

The main advantage of all the sections in Figure 11.2 is that they have greatly improved torsional stiffness. The closing of the section by the spot welds completes the path for the shear stress and reduces the warping effect. The cross-sections through the section remain near to a plane section. This results in the torsion constant or polar

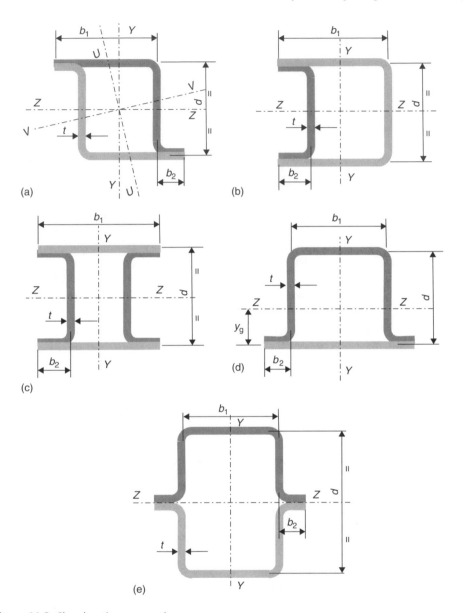

Figure 11.2 Closed sections–properties.

second moment of area being much greater and is given by (Gere and Timoshenko 1991):

$$J_x = \frac{4A^2 t}{s}$$

where A is the area enclosed by the median line, t is the wall thickness and s is the length of the periphery.

This formula really applies to a homogeneous section and in these sections consisting of several components the median line through the spot welds is taken as this is

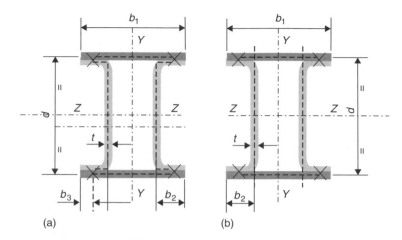

Figure 11.3 Enclosed area for evaluating the torsion constant.

effectively where the shear force is transferred between the components. Figure 11.3 shows this for the double channel closed with flat plates.

At (a) the enclosed area is:

$$A = (b_1 - 2b_2)d + 4(b_2 - b_3)t$$

And the periphery:

$$s = 2(b_1 - 2b_3) + 2d + 4(b_2 - b_3)$$

a more simple but less accurate value is shown at (b) where:

$$A = (b_1 - 2b_2)d$$

And
$$s = 2(b_1 - 2b_2) + 2d$$

This tends to give a slightly increased value for the torsion constant. From this analysis it is clear that for good torsional stiffness and strength it is essential that closed sections are used. The modern passenger car is constructed with most beams constructed as closed sections. The next section gives examples of typical closed sections used in car structures illustrating that designers have paid due attention to the desirable structural properties.

11.2.3   Passenger car sections

The previous paragraphs have considered idealized beam sections, but actual passenger car sections, although built up from angles, channels, Z sections and flat sheet, are rather more complex in detail profiles. Actual profiles are governed by numerous factors, such as styling lines, attachments for seals, attachments for glass, attachment for hinges, press tool draft angles, to name but a few. Figure 11.4 shows typical sections used in passenger car bodies. The importance of using closed sections discussed in the previous section is illustrated by the examples shown.

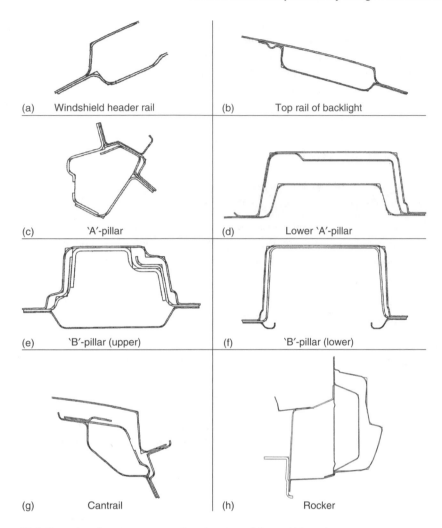

Figure 11.4 Examples of passenger car sections (courtesy of General Motors).

When making stress calculations, evaluation of section second moments of area and torsion constants can be evaluated using modern CAD methods. In choosing these sections the structural properties must always be borne in mind as well as the practical installation and functional requirements. As all the sections are closed sections the shear centre will be within the closed area and therefore the problem of local torsion is not significant. However, the use of closed sections does assist in achieving a higher torsional stiffness for the complete structure.

By examination of the real sections in Figure 11.4 it can be seen that the windscreen header rail (a) is made from a shallow hat section and the roof panel. The attachment under the roof is by bonding as the marks of spot welds on the outer roof surface are unacceptable. The top rail of the backlight (b) is constructed in a similar manner. The upper part of the 'A'-pillar shown at (c) is the edge of the windscreen and also the door frame. On one flange it has to accommodate the glass (bonded) and on the other the door seal. It is basically two hat sections. At (d) the lower 'A'-pillar is made of two hat sections and must provide supports for the door hinges. Sections for the upper and lower 'B'-pillar (e) and (f) are modified hat sections with approximately flat plates or very shallow hat sections forming the closing member. The lower section shown here has a large cut-out for access to items such as seat belt mechanisms. The cantrail shown at (g) consists of two shallow hat sections, an angle plus the roof panel. This has to provide not only structural strength to the sideframe but support for the door seal and a rain gutter. The sill or rocker (h) is made up of a number of hat, plate and Z sections to form a double box section.

11.3   Examples of initial section sizing

From the SSS analysis described in Chapter 5, the loads that are applied to members such as the floor cross-beams or the 'A'-pillar can be obtained. The next step is to determine suitable sections to carry these loads, taking into account any load factors (see Chapter 2, section 2.4).

11.3.1   Front floor cross-beam

As an example for determining the size of a structural member consider the floor cross-beam positioned under the front seats. This member will have loads in the vehicle bending case from the front passengers and seats, and may also have loads from the front longitudinals or engine rails. These loads are shown in Figure 11.5 where F_{fp} is the load from the front passenger/seat and K_1 the load from the engine rail. The cross-beam is often constructed from a hat section mounted on top of the floor panel and may also have a raised centre to pass over the equipment tunnel running longitudinally

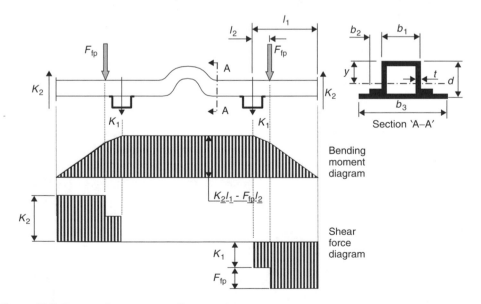

Figure 11.5 Forces and moments on a floor cross-beam.

down the passenger compartment. The loading condition on the cross-beam is bending and shear. The reactions K_2 are assumed to be simple support reactions with no fixing moment (the reason for this is given in section 11.4). With this assumption the bending moment and shear force diagrams shown may be obtained. There is no shear over the centre between the loads K_1 and the constant bending moment along this part is equal to:

$$M = K_2 l_1 - F_{fp} l_2$$

The values of K_2 and F_{fp} must be based on the static loads multiplied by any load factor necessary to allow for dynamic effects. When designing for strength use the standard engineer's bending theory to obtain the stress in the beam due to the bending moment:

$$f = \frac{My}{I}$$

where y is the distance from the section neutral axis to the top of the beam and I is the second moment of area of the section.

The section properties should be evaluated including a width of approximately $20t$ of the floor panel either side of the hat section (i.e. $b_3 = b_1 + 2b_2 + 40t$).The value of the bending stress under the action of the fully factored load should not exceed two-thirds of the material yield stress when using steel construction. This means that there is a safety factor against yield of the material under the worst load case of 1.5. This is also usually sufficient to prevent fatigue failures. From the shear force diagram it can be seen that the highest stresses will occur towards the ends of the beam where they are attached to the sills. The shear force on the beam here is K_2 and using the non-linear shear stress theory (Ryder 1969) the shear stress is:

$$\tau = K_2 A\bar{y}/zI$$

where A is the area of the section between the extreme edge and the point where stress is calculated, $\bar{y}$ is the moment of this area about the neutral axis, z is the width of the section where stress is calculated and I is the second moment of area about the neutral axis.

Using this formula shows that the stress in the top of the hat and in the floor panel is very low. Therefore, as a first approximation the shear stress may be assumed to be carried almost entirely in the sides of the hat section:

$$\tau \simeq K_2/2dt$$

Stiffness of the beam or deflection at the centre can also be considered in deciding the required second moment of area. Standard beam deflection theory such as Mohr moments area or Macaulay's method may be used (Ryder 1969).

11.3.2   The 'A'-pillar

The 'A'-pillar has large bending loads when the structure is loaded in torsion. From the roof loads the shear force between the top of the windscreen frame and along the cant-rail can be obtained. These in turn can be distributed into the sides of the windscreen ('A'-pillar) and to the pillars of the sideframe as shown in Chapter 7, section 7.2. If

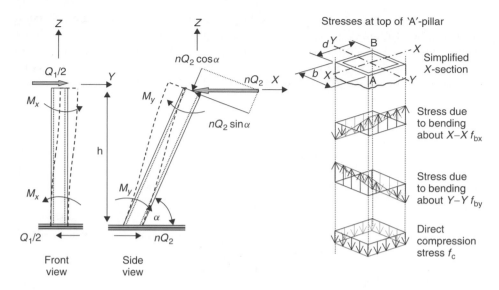

Figure 11.6 Forces, moments and stresses on 'A'-pillar.

the roof edge loads are Q_1 acting across the front and Q_2 acting along the sides, the loads on the 'A'-pillar are $Q_1/2$ in the Y direction, and nQ_2 in the X-direction (see Figure 11.6). The proportion n of the sideframe load Q_2 is that taken by the 'A'-pillar. Assuming the joint at the top of the 'A'-pillar to the windscreen header rail/cantrail and the joint to the dash/wing are 'encastré' then the pillar will deflect as shown in Figure 11.6. Bending moments occur at the top and bottom. As seen in the front view the bending moment at the top acting about the $X-X$ axis is:

$$M_x = Q_1 h/4$$

This will cause a bending stress:

$$f_{bx} = \frac{M_x b}{I_{xx}2}$$

In side view the force nQ_2 can be resolved into $nQ_2 \sin\alpha$, α acting normal to the axis of the pillar, and $nQ_2 \cos\alpha$, α acting along the axis. The normal force gives rise to a bending moment:

$$M_y = (nQ_2 \sin\alpha)\, h/2 \cos\alpha$$

and this will cause a bending stress:

$$f_{by} = \frac{M_y d}{I_{yy}2}$$

The direct force will put compression into the pillar giving a stress:

$$f_c = \frac{nQ_2 \cos\alpha}{(2b + 2d)\,t}$$

where $(2b + 2d)t$ is approximately the cross-sectional area.

The resultant stress on the 'A'-pillar is the sum of the direct compressive stress f_c, the bending stress f_{by} and the bending stress f_{bx}. Examination of the stress plots shown in Figure 11.6 shows that at corner B all three stresses are compressive which gives the maximum stress condition. At corner A the bending stresses are tensile but the direct stress is again compressive so giving a reduced resultant stress.

The design criteria described in this section are based on the torsion load condition for the vehicle, including any dynamic load factors (see Chapter 2). The criteria for choosing the 'A'-pillar section can, however, be the in-roof crush test SAE J374. Structural crashworthiness is, however, beyond the scope of this book.

11.3.3   Engine longitudinal rail

This component as shown in Figure 11.7 is subject to shear forces and bending moments generated from a number of loads. These loads are caused by the weight

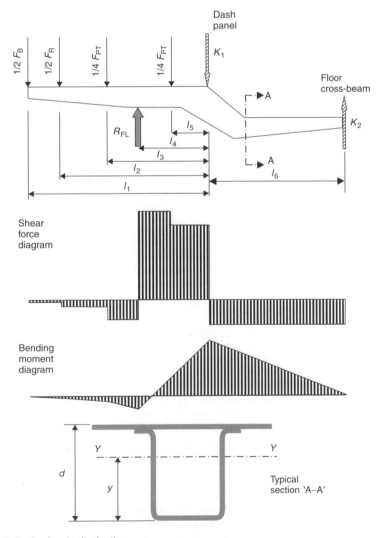

Figure 11.7 Engine longitudinal rail.

of components such as the bumper F_B, the radiator F_R, the power-train F_{PT} and the reaction from the front suspension R_{FL}. The weights, of course, are shared between two rails but need to be factored by a dynamic factor to allow for the effects of striking kerbs, pot-holes, etc. (see Chapter 2).

The engine rail is supported in the structure by the dash panel and by the floor cross-beam situated under the front seats. The forces K_1 and K_2 can be obtained from the equations:

Resolve vertically:

$$\tfrac{1}{2}F_B + \tfrac{1}{2}F_R + 2(\tfrac{1}{4}F_{PT}) + K_1 - K_2 - R_{FL} = 0$$

Moments about K_1:

$$\tfrac{1}{2}F_B l_1 + \tfrac{1}{2}F_R l_2 + \tfrac{1}{4}F_{PT}l_3 + \tfrac{1}{4}F_{PT}l_5 + K_2 l_6 - R_{FL}l_4 = 0$$

The plots of shear force and bending moments along the beam show high shear between the suspension reaction and the dash panel and the maximum moment is at the dash panel. Hence, the need for a deeper section in this area is demonstrated. Usually the design is governed by the bending strength requirement at the dash panel, hence the overall depth d should be large. Stiffness may also be important and the deflection of the beam calculated by any of the recognized methods. This member will also be designed to absorb energy in frontal impacts but this requirement is outside the scope of this book.

11.4 Sheet metal joints

There are many spot welded joints in a passenger car body and one example is shown in Figure 11.8. This is typical of the floor cross-beam to sill joint and it is an example from which to learn some basic principles. In section 11.3.1 it was stated that this joint has no fixing moment to the sill providing only a shear connection carrying the vertical load. Figure 11.8(b) illustrates why this condition exists. If a moment M is applied it will cause tension in the top of the hat and compression in the closing plate (the floor panel). These loads will cause the top flange to unfold putting tension into the spot welds, which is also undesirable. The flange of the lower plate (floor) will tend to buckle under compression while the side flanges will twist. The net result is there cannot be any significant bending stiffness.

Vertical shear force is carried by the side flanges, which act like curved shear panels, but the top and bottom flanges will not carry any significant vertical force as shown in Figure 11.8(c). If the side flanges are removed the top of the hat and the floor panel will act like a bridge between the sill and cross-beam. As they are thin plates and the load is applied normal to their plane they will be ineffective in resisting the force and will distort as shown.

When horizontal shear forces are applied the opposite will result. That is the top and bottom flanges will carry the horizontal force while the side flanges will be ineffective.

The spot welds on the side flanges alone must be designed to carry the vertical loads and those on the top and bottom to carry the horizontal loads.

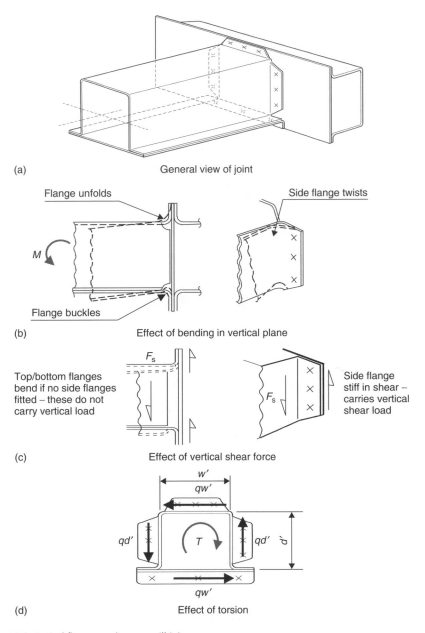

(a)   General view of joint

(b)   Effect of bending in vertical plane

(c)   Effect of vertical shear force

(d)   Effect of torsion

Figure 11.8 Typical floor cross-beam to sill joint.

If the cross-beam is subject to a torsion moment as shown in Figure 11.8(d) the shear flow around the section can be obtained from:

$$q = \frac{T}{2A} = \frac{T}{2d'w'}$$

Therefore the shear force on the vertical sides will be qd' and along the top and bottom qw' hence the shear force acting on the spot welds is approximately equal to these values. This gives a conservative result as the spot welds are spaced further from the beam centre and hence the load will be reduced.

From the analysis of this joint we can learn two important rules for designing joints:

1. Avoid out-of-plane bending on thin sections.
2. Load thin sections with in-plane bending and shear.

11.4.1 Spot welds

In the previous section it was noted that loading spot welds in tension was unsatisfactory. Figure 11.9 illustrates the problems of loading in tension or twisting and the benefits of providing a direct shear load. At (a) there is a section through a spot weld

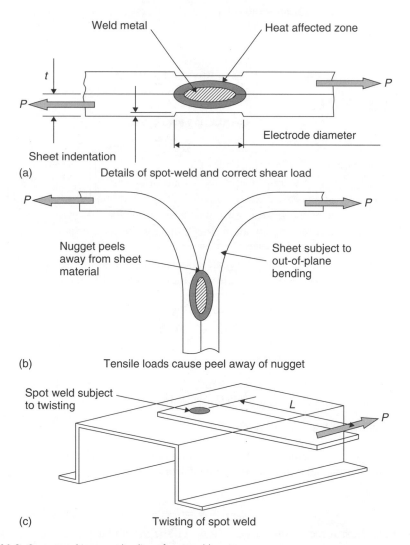

(a)　　　　Details of spot-weld and correct shear load

(b)　　　　Tensile loads cause peel away of nugget

(c)　　　　Twisting of spot weld

Figure 11.9 Correct and incorrect loading of spot welds.

showing a centre core which has the microstructure similar to a casting, although there is some working of the metal due to the pressure of the electrodes. Surrounding the core is a heat-affected zone that has reduced strength compared to the base material. Failure occurs in this zone as shown at (b) where the spot weld nugget has peeled away from the base material and at (c) where the small area cannot resist a large twisting moment caused by the long moment arm. Note also at (b) how the parent metal is subject to out-of-plane bending which again is unsatisfactory causing yielding at very low loads.

11.5   Spot weld and connector patterns

Suspension arms and other components are often attached to brackets that in turn are attached to body structural beams with a group of spot welds as shown in Figure 11.10. At (a) of this diagram, it may be observed that the load P is offset from the spot weld groups with a moment arm e from the centroid.

The centroid of the group can be found by taking moments about a horizontal line to obtain $\bar{y}$ and about a vertical line to obtain $\bar{x}$. For this example:

$$\bar{y} = (3y_1 + 3y_2)/8$$
$$\bar{x} = (2x_1 + 3x_2)/8$$

The offset load P can now be replaced by a load P acting through the centroid and a moment $M = Pe$ about the centroid shown at (b). Assuming the plates remain rigid then under the action of the load P through the centroid the distortion and hence the load on each spot weld will be equal. In this example the load on any spot weld n is $F_{Cn} = P/8$.

The moment will tend to rotate the plate about the centroid and again assuming the plate is rigid the strain on each spot weld will be proportional to the distance from the centroid:

$$\frac{F_{E1}}{r_1} = \frac{F_{E2}}{r_2} \dots \frac{F_{En}}{r_n} = K$$

Taking moments about the centroid:

$$\sum_1^n F_{En} r_n = M$$

Therefore:

$$\sum_1^n K r_n^2 = M$$

$$K = \frac{M}{\sum\limits_1^n r_n^2}$$

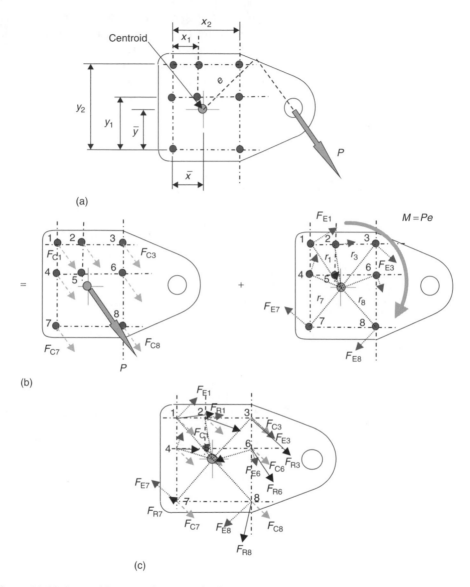

Figure 11.10 Spot weld group with eccentric loading.

and
$$F_{En} = \frac{M r_n}{\sum\limits_{1}^{n} r_n^2}$$

These forces act at right angles to the line from the particular spot weld to the centroid.

Taking the vector sum of the forces F_{Cn} and F_{En} the resultant force on each spot weld is obtained as shown in Figure 11.10(c). Note in this example spot weld 3 is the most heavily loaded while spot weld 7 has very little load. There will always be some spot welds that will have very reduced load, which means that some repositioning could be made so that a more even load distribution results. Practical requirements, however, may prevent this.

11.5.1   Spot welds along a closed section

The box section shown in Figure 11.11(a) consisting of two channel sections and two plates has four rows of spot welds joining the parts together. Consider first, a torsion

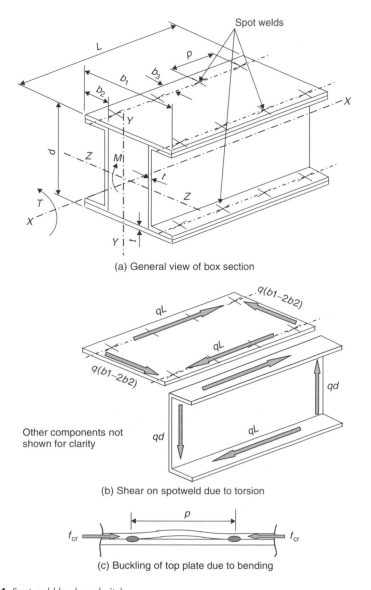

(a) General view of box section

(b) Shear on spotweld due to torsion

Other components not shown for clarity

(c) Buckling of top plate due to bending

Figure 11.11 Spot weld loads and pitch.

moment T applied about the longitudinal axis $X-X$. This causes shear stress τ to all four components, which can be evaluated from:

$$\tau = \frac{q}{t} = \frac{T}{2At} = \frac{T}{2d(b_1 - 2b_2)t} \quad \text{(approximately)}$$

where q is the shear flow (force per unit length), t is the wall thickness and is assumed to be the same for both channels and plates, A is the enclosed area of the box, d is the median depth between the top and bottom plates, b_1 is the overall width and b_2 is the width of the flange of the channels.

The shear force along one row of spot welds will be qL (see Figure 11.11(b)). If the strength of the spots is F_s the number of spots required in each row is:

$$N = \frac{qL}{F_s}$$

and the pitch p between the welds can easily be obtained. The end spot welds should not be placed less than twice the electrode diameter from the ends.

For typical box sections used in passenger cars the number of spot welds required to meet this torsion load is relatively small and therefore the pitch is large. This means that the assembly of channels and plates departs considerably from a homogeneous structure and from the concept of a constant shear stress assumed in the thin walled theory. Another feature is that when the pitch between spot welds is large this can lead to buckling problems for the bending load case. When a bending moment M is applied about the $Z-Z$ axis as shown in Figure 11.11(a) the top plate is loaded in compression. As the plate is thin (t is approximately 1 mm for car sections) and the pitch p is large, the plate will act like a strut (see Figure 11.11(c)). The spot welds provide some clamping effect but do not provide a fully encastré condition.

From classical Euler strut theory the buckling load for a fully encastré strut is:

$$P_e = K\pi^2 EI/L^2 \quad \text{(Ryder 1969)}$$

where E is the Elastic Modulus, I is the second moment of area, L is the length of the strut and $K = 4$ for fully encastré.

From ESDU 02.01.08 the clamping effect of a spot weld is such that $K = 3.5$, and the buckling stress f_{cr} for a thin plate between rivets can be found from:

$$f_{cr} = \frac{P_e}{A} = 3.5\pi^2 EAk^2/Ap^2$$

where E is the modulus of elasticity, A is the cross-sectional area of the plate, k is the radius of gyration $= t/2\sqrt{3}$ for a plate of thickness t and p is the pitch of the spot welds.

The actual stress applied due to the bending moment is:

$$f_b = \frac{Md}{2I}$$

which must not be greater than the buckling stress f_{cr} or alternatively the spot weld pitch must be less than:

$$p = \left(\frac{3.5E\pi^2 t^2}{12 f_b} \right)^{1/2}$$

The pitch requirement to prevent buckling between the spot welds is usually the design criterion for this type of box beam as used in passenger cars.

11.6   Shear panels

In Chapters 4 and 5 it was found that the floor panel, dash panel, roof, inner wing panel, etc. are loaded in shear. When designing these panels it should be realized that the shear loading creates a tension field, which runs approximately diagonally across the panel. At right angles to this there is a compressive field which can lead to buckling of the panel. Fortunately in passenger car panels the shear stress is relatively low but the panels tend to be quite large which in turn leads to a low buckling stress.

11.6.1   Roof panels

The roof panel is probably the largest panel in a passenger car and under the torsion load case this may buckle due to shear. ESDU 02.03.18/19 data sheet can be used to investigate this phenomenon. These data is not ideal because they consider plates with curvature in one direction, but the roof panel has curvature in two directions. Nevertheless using these data the shear stress at which the panel buckles is given by:

$$\tau = KE(t/b)^2$$

where K is the buckling stress coefficient, E is modulus of elasticity, t is the panel thickness and b is the length of the curved side.

The coefficient K is presented as a function of the length a, the radius of curvature R, the thickness of the panel t and the ratio a/b (the ratio of the lengths of the panel sides).

Investigations made into a roof panel 980 mm wide by 1250 mm long, 1 mm thick and radius of curvature of 2425 mm resulted in stress to cause buckling of 13.8 N/mm^2 which is four times larger than the applied shear stress.

A more conservative investigation is to consider the roof as a flat panel (which agrees with the concept of a simple structural surface) and use ESDU 71005. A similar equation is used to evaluate the buckling stress as that used for the curved panel. In this case b is the length of the shorter side and K is presented as a function of the ratio of the sides b/a and the edge condition. Edge conditions given vary from simple support on all edges to fully clamped on all edges with all the alternatives in between. Results using this information for a similar panel to that mentioned in the previous paragraph give buckling shear stress as 1.5 N/mm^2 (simple supported edges) to 2.4 N/mm^2 (fully clamped). These are both less than the applied shear stress indicating the possibility of buckling. Placing a stiffener across the panel parallel to the shorter side and midway along the length increases the respective buckling stresses to $3.3\,\text{N/mm}^2$ and $5.4\,\text{N/mm}^2$. These are both higher than the applied shear stresses.

From these investigations it can be observed that roof panels will buckle at low stress levels but with the addition of curvature and a central stiffener across the roof the buckling stress can be raised well above the applied stress.

Another approach to designing roof and other panels is to consider the vibration characteristics. When riding in a commercial vehicle such as a light van the low frequency drumming of roof and side panels is abundantly manifest. ESDU 75030 can be used to predict panel vibrations. Returning to the roof panel previously described and using this data sheet the lowest fundamental frequency is approximately $4\,\text{Hz}$ with various mode frequencies between this value and $33\,\text{Hz}$.

This reveals that low frequency vibrations will cause ride and comfort problems and that the stiffening and clamping of roof panels are more important than shear buckling.

11.6.2   Inner wing panels (inner fender)

Using the SSS model that was considered in Figure 5.1 the shear stress on the inner wing panel between the suspension mounting and the dash panel can be significant, especially when the bending and torsion cases are combined. Also the pot-hole loading must be investigated and may well be even more severe.

A typical panel may well be modelled with dimensions of $640\,\text{mm}$ high by $480\,\text{mm}$ long. Once again using ESDU 71005 buckling stresses are $6.1\,\text{N/mm}^2$ and $9.8\,\text{N/mm}^2$ for simply supported and clamped edges respectively. Applied stress levels may exceed these values by a factor of 5, hence there is the need to provide stiffeners to prevent buckling.

Fortunately this panel in practice has considerable curvature, is restrained at the edges by adjacent parts and also the model is a very simplified representation of the structure. In practice the load will be shared between the engine rail, the fender top rail as well as the panel. All these factors will tend to reduce the risk of panel buckling but this does illustrate the need to add stiffeners and swages in this part of the structure.

Case studies – preliminary positioning and sizing of major car components

12.1   Introduction

In this chapter the concepts from previous chapters will be utilized to illustrate how the SSS method may be applied to impact early concept design decisions. These are presented as hypothetical examples to aid understanding and are not specific to any company's particular vehicle design. The solutions proposed are among many possible alternatives to vehicle body design and should not be construed as rigid guidelines.

12.2   Platform concept

At the early stage of the vehicle concept process, a high level description of the customer wants and needs will be explored. Very often these aren't yet fully translated into engineering requirements. At the same time, there will be a need to rapidly assess which platform(s) are the best candidates for the foundation of the new vehicle. Indeed, more than one vehicle type may be under consideration to compete in existing market segments or to develop a new market segment. Questions will be asked such as: 'What available platform(s) best conform to the customers' wants and needs?', 'What are the manufacturing constraints in building such a vehicle off this platform(s)?' Each candidate platform must be considered for production volume capacity, flexibility of the process to produce the vehicle type(s) in question, etc. The results of this investigation will converge to one or two primary candidate platforms, with the possibility that a new platform may have to be developed.

At this point it is perhaps useful to revisit the definition of 'platform'. There is no single definition that is standard across the industry. For the purpose of this book, a platform is defined as the assembly of parts which form the primary body front end, dash, floor, lower rear end and chassis. A platform may be lengthened, widened, or shortened and still be designated as the same platform. There may be specific parts for, say, the rear compartment pan of one variant that may not be common with others. Sometimes a platform is defined in terms of manufacturing constraints (i.e. if all of the model variants can be built in the same plant). A platform may be designated as the primary structural load path for the vehicle. Upon a given platform there may be multiple body types and sizes. It is assumed here that the model variants are represented by unique sideframes, although this is a general simplification. The collection of structural elements making up the specific body variant may be considered to be the secondary load path. Thus a platform whose lead body was a sedan may also produce (but not be necessarily limited to) a station wagon/estate car, hatchback and coupé. A platform whose lead vehicle was a van may produce an extended wheelbase van, a station wagon/estate car, and a pick-up truck. Within each platform may be the capability to support a bandwidth of power-trains, suspensions, and tyres. It is not given or necessarily desirable that a platform share 100 per cent common parts across all its model variants. Very often some new and variant-specific parts will be needed to accommodate a particular model's functional, aesthetic, and manufacturing requirements. Thus a platform should be regarded as a flexible entity, not frozen. For example, Figure 12.1 shows seven Volkswagen model variants (which may also be called derivatives) from the same platform.

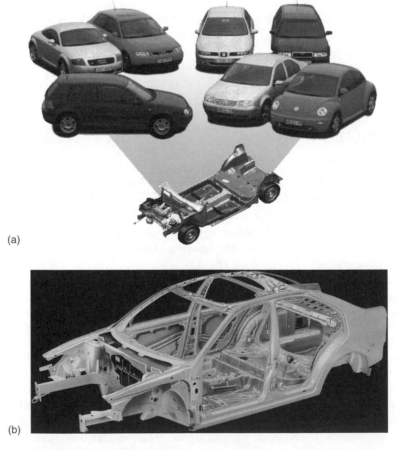

(a)

(b)

Figure 12.1 (a) Volkswagen platform derivatives (courtesy Volkswagen AG); (b) Volkswagen Bora body structure (courtesy of Volkswagen AG).

3-door hatchback

Saloon and
5-door hatchback

Wagon

Monocab

Figure 12.2 GM Opel platform derivative models.

In another example, Figure 12.2 shows five GM Opel model variants from a given platform.

Each platform derivative model may be differentiated by a specific character, and aimed at possibly a different market segment. Depending on many factors, a platform's structural members may or may not have been originally sized to accommodate all future potential variants.

12.3 Factors affecting platform capability for new model variants

The following factors under consideration for new body variants will change the magnitude of the edge loads (Q), internal reaction loads (P), bending moments (M) and torques on a platform structure. A determination must be made as to the capability of the platform to support newly proposed model variants:

1. *Weight.* The new vehicle may be projected to have a gross vehicle weight (GVW) higher than the lead body because of size, payload, power-train selection, etc. These will produce higher fundamental loading at the suspension reaction points (Rp and Rr) and higher bending moments.
2. *Vehicle type.* There may be a shift in weight distribution over the axles if, say, the new variant is a utility vehicle such as a pick-up truck. Typically on such vehicles a higher percentage of the payload weight is carried over the rear axle, so the suspension loads there will be higher. This will also shift the location of the maximum bending moment. A change in the CG (centre of gravity) location must also be considered in weight transfer calculations under braking and cornering

conditions. If the new variant is a convertible, the floor pan structure of the base platform may have to be modified, as shown in Chapter 6, section 6.4.

3. *Duty cycle.* The new variant may be targeted for a region or part of the world where the durability requirements or duty cycles are more severe because of road conditions or other factors. If the new variant is intended to be a pick-up truck or sport utility vehicle, then a different duty cycle may be required to account for customer use, such as off-road. Thus higher structural fatigue damage factors and/or peak load factors may be indicated.

4. *Dimensions.* Changes in track width, wheelbase, overall width, overall height, and overhang dimensions have the potential to change the values of the internal member forces, edge forces, and bending moments. This has been demonstrated in previous chapters.

5. *Suspension.* Changes in tyre size and profile, suspension travel, suspension type, and ride tuning parameters can affect the magnitude of the fundamental loads into the body structure. For instance, reduced suspension travel over bumps can result in higher loads into the body structure. Similarly, changes in shock absorber damping characteristics can influence peak loads during severe pot-hole events. Lower profile tyres will tend to increase the stiffness of the tyre sidewall and impart higher loads into the body. Larger wheels and tyres will increase the unsprung mass and therefore the suspension loads into the body.

In the conceptual design stage, there is limited time to do detailed studies on all the above. Exhaustive detailed studies may come at the expense of only being able to consider one or two alternatives. In today's highly competitive market place, there is a need to rapidly assess many different alternatives and market segment opportunities.

12.4   Examples illustrating role of SSS method

Examples of how the SSS method can offer a preliminary coarse assessment regarding items (1), (2), and (4) above, are explained below.

12.4.1   Weight

All other things being equal, a change in total vehicle weight will mean a linear increase in the wheel load reactions and suspension inputs. The edge forces, internal reactions, and bending moments of the SSS model will change proportionately. These can then be checked against the platform's structural member capability to support the loads using hand calculations from engineering mechanics and finite element analysis as appropriate. This capability could be expressed, for example, in terms of cross-section bending moment capacity to resist permanent deformation or its stiffness to resist deflection. For example, the longitudinal engine rails depicted in Figure 8.5 are assumed to have a defined length, cross-sectional shape, thickness, material yield strength, modulus of elasticity, and moment of inertia (second moment of area). For simplicity of the method, the member is idealized as an SSS. Previous load–shear–bending moment diagram plots based on engine and suspension loads are shown in Figure 8.7. From this information the maximum *allowable* bending moment and its location can be calculated assuming yield stress criteria multiplied by an appropriate safety factor. If deflection is the governing criterion, the displacement can be calculated from simple beam formulae, again multiplied by a safety factor. An increase in total vehicle and power-train weight for the proposed model variant will produce proportionately higher suspension and engine mount loads on the rail member, respectively. Bending moments can then be recalculated and compared to an *allowable* bending moment capacity. Beam deflection could then be calculated and compared to a target value established by experience or from a reference vehicle.

12.4.2 Vehicle type

It has been shown in previous chapters that changing the vehicle body type can change significantly the resolution of the edge forces and load reactions within the structure. These changes affect the magnitude and shape function of bending moments and shear forces on parts that may have erroneously been assumed to be unaffected.

The following are presented as general hypothetical examples to illustrate the method and are not specific to any company's particular vehicle design.

12.4.3 Sedan to station wagon/estate car – rear floor cross-member

Take for a first example the sedan shown in Figure 8.1. It is assumed the underfloor rear longitudinal rails, rear suspension cross-member and trunk floor may be carried over to create a station wagon/estate car model variant. The station wagon/estate car is shown in Figure 4.12. In both cases the rear floor cross-member serves as a means of supporting the cargo weight and transferring load to the sideframe. However, the payload rating for the wagon was selected on the basis of being competitive with wagons of similar class, with the result that the rear cross-member will see higher loading than originally designed for the sedan. Shear and bending moment diagrams of the cross-member conceptualized as an SSS member (as in Figure 8.11) can be used as a precedent to calculating the allowable bending moment under stress and/or deflection limit criteria.

12.4.4 Closed structure to convertible

Another example is the modification of a closed structure to create an open convertible model, as described in Chapter 6, section 6.4. Of the fundamental load cases being considered in this book, torsion is the most extreme condition for the platform structure. Convertibles tend to be produced in relatively low production volumes compared to other model variants. It is typically difficult to justify the cost and packaging compromises of incorporating the additional reinforcing structure into the base platform. Thus the remedies for the open structure are often made specific to the convertible model only. This is illustrated in a portion of the body described in Figure 12.3.

The original body was a closed structure. Finite element analysis showed that removing the roof structure reduced the torsional stiffness from 12 000 N-m/deg. to

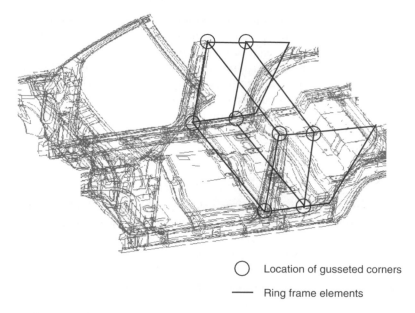

○   Location of gusseted corners

—   Ring frame elements

Figure 12.3 Illustration of ring frame concept on an open body structure to increase torsional stiffness.

1700 N-m/deg. Referring to Chapter 6, section 6.4, it was shown that the incorporation of SSS boxes to generate shear flow can be an effective means of recovering some of the torsional rigidity. These were simulated in a finite element model using multiple and integrated ring frames with gusseted corners, the locations of which are shown in Figure 12.3. Frame section sizes ranged from $80 \times 100 \times 1.5$ mm to $120 \times 120 \times 1$ mm. These measures alone increased the torsional stiffness by 2000 to 2500 N-m/deg. Other significant improvements which approximated the fundamental countermeasures illustrated in Chapter 6, section 6.4 were (1) a structural tunnel (integral with the #1 bar and the rear ring frame), and (2) longitudinal members reinforcing the floor pan (from the dash and #1 bar to the #3 bar) to form a grillage. The incorporation of the above countermeasures together were significant in improving the torsional stiffness by over a factor of three times the baseline open structure. It should be pointed out that additional structure alone will be ineffectual without fundamentally continuous and well-distributed load paths from the engine compartment rails/front suspension to the rear longitudinal rails/rear suspension load points.

12.4.5 Dimensions

This example uses wheel track. Recalling Chapter 2, section 2.4.2, an increase in the wheel track will increase the torsion load on the vehicle structure, assuming all other factors remain unchanged. The increase in wheel track may be accompanied by a proportionate increase in vehicle width and therefore the cross-car dimensions of the dash, floor, parcel shelf, and cross-members. Reviewing the free body diagrams in Chapter 5, section 5.3 and solving for the equations in Chapter 5, section 5.3.5 can give a first order assessment of the impacts. The effects of increasing the width dimension on the edge forces can be quantified by resolving the equations derived from previous chapters. The revised edge forces can then be compared with a baseline reference vehicle body.

The above examples concentrated mainly on the vehicle underbody platform. The same principles can be applied to assess the impacts of the above factors on structures that are specific to the body type, such as the sideframe, roof, and rear seat back support.

The next illustration will be to explore the bandwidth of potential vehicle design variants from a given platform.

12.5 Proposal for new body variants from an existing platform

In this hypothetical case study, it is proposed to develop a mid-sized rear wheel drive pick-up truck, a mid-sized van, and a station wagon (estate car) from a platform originally designed for a smaller existing production mini-van. SSS idealizations are shown in Figure 12.4.

This example problem is intended to illustrate the process of applying the SSS method with engineering fundamentals. The *original* assumptions are:

1. Body frame integral (BFI) or unit-body construction is to be retained.
2. Front end, centre floor, and rear floor structures are to be carried over from the mini-van except for length modifications.
3. The front wheels will be moved forward by a specified amount only on the variant models to allow room for a larger engine and styling changes. The wheelbase will be extended for the mid-size van by a specific amount and even further on the pick-up truck, respectively.

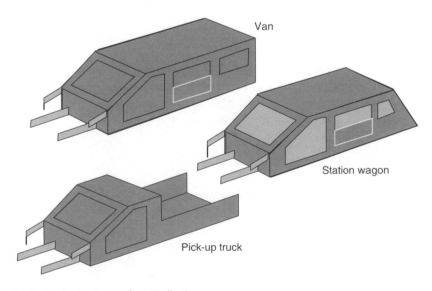

Figure 12.4 Simple structure surface idealizations.

4. Overall width and height will remain similar to the mini-van. The front wheel track will be widened somewhat, which will also extend the lateral distance between front suspension load points by the same amount.
5. Front and rear suspensions will be carried over from the mini-van.
6. Total payload will be constrained to the mini-van's capacity, with the exception of the mid-size van that will be increased.
7. Front end, rear end, and floor structure designs will be carried over from the mini-van (except for modifications required to lengthen the wheelbase). The sideframes, roofs, rear liftgate opening frames, windshield frames, rear seat bulkhead (truck), and rear cargo box (truck) will be all new. The windshield will be moved forward on the pick-up truck, essentially eliminating the cowl as an intermediate structure panel.

The next step is to translate the above assumptions into working definitions for first-order SSS studies. The *torsion* load case will be discussed.

12.5.1   Front end structure

There is a motor compartment lower midrail, an upper motor compartment rail, and a shock or strut tower vertically connecting the two. Idealization of this structure is similar to the example shown in Figure 6.12. This arrangement is shown is carried over in Figure 12.5.

The shock cap vertical suspension load is split between the lower midrail and the upper rail. The upper rail is in effect a cantilevered beam whose bending moment and vertical reaction force are reacted at the sideframe/FBHP (front body hinge pillar). The upper rail and midrail members are correspondingly extended from the dash panel on all the variant models.

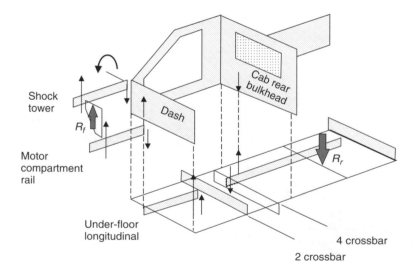

Figure 12.5 Underbody load-path assumptions for pick-up truck (right side shown).

12.5.2   Dash

This panel will react the vertical load from the midrail underfloor longitudinal member and will be balanced by edge forces at the front body hinge pillar (FBHP), windshield frame/cowl, and floorpan. This is illustrated in Figure 12.6.

12.5.3   Floor

The van and station wagon are differentiated from the pick-up truck in how the rear suspension loads are reacted by the body structure. This is illustrated in Figures 12.7 and 12.8.

The floor in reality has multiple cross-members that make the reactions statically indeterminate, so the problem must first be reduced to the minimum number required for static equilibrium. The front suspension vertical loads distributed to the midrails are reacted by the dash and the #2 underseat cross-member. The midrail longitudinal-to-cross-member vertical reaction loads are equal, but are opposite sign left to right. Both cross-member loads are reacted by the sideframes. The rear suspension vertical loads on the station wagon and van longitudinal rails are reacted by the rear liftgate opening ring at the back and the #5 floor cross-member (or rear seat riser) at the front. The #5 cross-member is idealized as the location where floorpans between the van/station wagon and the pick-up truck are differentiated. The floorpan also rises at this point to accommodate the rear cargo floor. It is also at this location where the pick-up truck cab

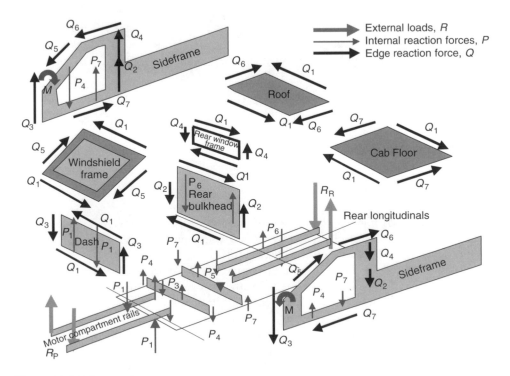

Figure 12.6 Pick-up truck.

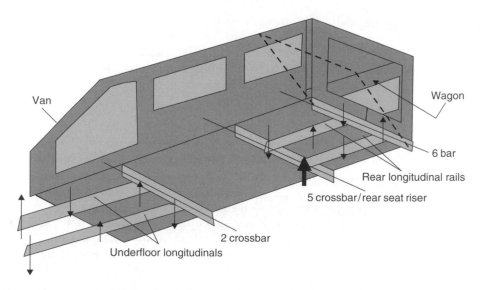

Figure 12.7 Rear wheel drive van/station wagon underbody load-path assumptions.

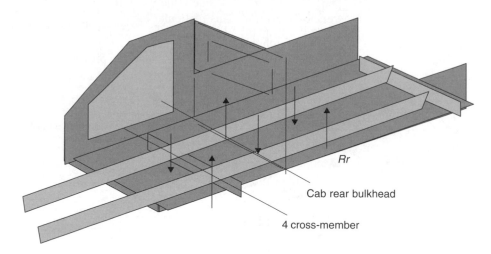

Figure 12.8 Rear wheel drive pick-up truck rear load-path assumptions.

rear bulkhead is positioned. The #5 rear seat riser cross-member vertical loads from the rear longitudinal rails are reacted by the sideframe on the van/station wagon, and at the crew cab rear bulkhead on the pick-up truck.

The rear longitudinal rail is not sufficiently supported by the #5 rear seat riser cross-member on the pick-up truck. Unlike the van and station wagon rails which are simply supported, the pick-up truck does not have a rear end opening ring frame structure to support the end of the rear longitudinal rail.

Essentially it becomes a cantilevered beam. Even if a stiff floor cross-member was provided at the end of the longitudinal rails, it would see little support from the rear edge of the cargo box which is relatively narrow to accommodate the wide tail-gate opening. An alternative load path must be provided. This is accommodated by extending the rear longitudinal rail forward, past the #5 rear seat riser cross-member to the #4 floor cross-member under the front edge of the rear seat. The #5 floor rear seat riser cross-member is considered part of the cab rear bulkhead structure. Thus this load path becomes similar to how the front midrail forces are reacted. The consequences of this arrangement on the rear structure will become apparent when the forces are calculated and compared across the model variants.

12.5.4   Cab rear bulkhead (pick-up truck)

This structure is idealized as an SSS panel capable of supporting in-plane edge forces and also loads from the #5 underfloor cross-member. It consists of a fixed rear window frame (upper) and a sheet metal panel (lower), with support from the sideframe (cab side), roof, and floor. This is illustrated in Figure 12.6.

12.5.5   Sideframe and cargo box side

Because the cargo box is open, the pick-up truck sideframe's contribution to the body torsional stiffness is just the enclosed portion. This particular idealization assumes that none of the rear longitudinal rail vertical suspension loads are transmitted to the cargo box side panel. Therefore, there are initially no horizontal 'boom' forces to load the B-pillar from the cargo box as shown in Figure 12.9.

12.5.6   Rear compartment pan and cargo box floor

Only the van and station wagon can provide reactions to support the in-plane edge forces. On the pick-up truck, the cargo box floor is considered redundant under body torsional loading. Instead, the cab floor provides these forces as shown in Figure 12.6.

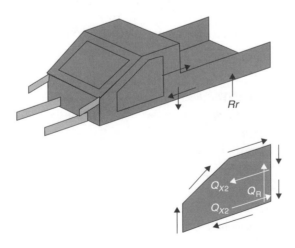

Figure 12.9 Alternative load path for pick-up truck.

The cargo box floor does, however, provide a close-out panel to stabilize and enclose the rear longitudinal rail hat cross-section.

This early visualization has already produced some insight into potential issues prior to the commencement of the SSS calculations: (1) the front end structure is extended from the dash which will produce larger bending moments on the upper rail and front body hinge pillar (FBHP) sideframe joint; (2) the assumption of common rear longitudinal rails differentiated only by length may have to be reconsidered in light of the pick-up truck load paths.

12.5.7   Steps for preliminary sizing of components

Step 1   Draw free body diagrams of each model variant for all the structural members. Figure 12.6 shows the pick-up truck example.

Step 2   Resolve the free body diagrams into equations and arrange these equations into matrix form. Figure 12.10 shows example equations for the pick-up truck. The matrices for each model variant are developed as shown in Figures 12.11 and 12.12.

Figures 12.11 and 12.12 also show the names and definitions for each of the input parameters arranged in spreadsheet form. The spreadsheet can be arranged such that the values of the input parameters can be copied into the matrix to be solved. In addition, a spreadsheet format can also facilitate the calculation and display of the non-edge force internal reactions (P, M) and the external reaction forces (Rr). It is also suggested that a separate table be set up to organize all the dimensional parameters for each model variant.

Step 3   Solve for the internal reaction forces P. These calculations are pre-requisite to solving for the edge forces (Q). This may be accomplished using a built-in spreadsheet program or hand calculator. For this problem, a unit value of 1000 N has been selected as the external applied load. This makes it easier to scale and compare results later.

Step 4   Solve the matrix equations to obtain the edge reaction forces Q. This may be accomplished using a spreadsheet program, scientific hand calculator, or other methods.

$$-Q_1L + Q_7B = 0 \text{ ... Floorpan}$$
$$-Q_1h_4 - Q_2B = P_6\,xr - P_6(B-xr) \text{ ...Rear bulkhead}$$
$$Q_1h_1 + Q_2B = -P_1f + P_1(B-f) \text{ ... Front dash bulkhead}$$
$$-Q_1h_3 + Q_4B = 0 \text{ ... Rear window frame}$$
$$Q_1h_2 - Q_5B = 0 \text{ ... Windshield frame}$$
$$Q_1l - Q_6 = 0 \text{ ... Roof}$$
$$Q_2L - Q_4\,L - Q_6\,(H-h_1) - Q_7\,h_1 = -M - (P_4\,a_1) + (P_7\,(L-a_2))$$
$$\text{... Cab sideframe}$$

Figure 12.10 Pick-up truck equations.

Matrix for coefficients:

	1	2	3	4	5	6	7	8	9	SOLVED Q EDGE FORCE (N)	CALCULATED INTERNAL LOADS	MATRIX EQUATION PART NAME	Matrix equ. edge force
1	c	-w	0	0	0	0	0	0	0			cowl panel	Q1
2	h-d/cos(alpha)	0	-w	0	0	0	0	0	0			windshield frame	Q2
3	g	0	0	-w	0	0	0	0	0			roof	Q3
4	k	0	0	0	-w	0	0	0	0			front floorpan	Q4
5	n	0	0	0	0	w	0	0	0			rr crossbar/rearseat riser	Q5
6	h-n	0	0	0	0	0	w	0	0			liftgate frame	Q6
7	p	0	0	0	0	0	0	w	0			rear compartment pan	Q7
8	d	0	0	0	0	0	0	w	w			dash	Q8
9	0	0	0	-(h-d)	-d	k-c	k+i+m-c	-(d-n)	c			sideframe	Q9

Parameter definitions:

name	value	definition
a (mm) long.		strut cap to sideframe and fod
d (mm) vert.		FBHP lnr lwr height
k (mm) long.		fod to rr seat riser
n (mm) vert.		rr seat riser depth
tr (mm) lat.		span between CL's of rr long rail loads
alpha (deg) from vert.		w/s inclination angle
Rf (N) vert.		strut cap unit load
M (N-mm) X-Z		Moment: P1 X a
b (mm) long.		fod to 2 Xbar
g (mm) long.		roof length
l (mm) long.		rr seat riser to rr long rail load pt
p (mm) long.		rr seat riser to rrend pnl
w (mm) lat.		width of roof
P7 (N) vert.		rr seat riser @ rr long. rail load
Rr (N) vert.		rr long rail susp load
P4 (N) vert.		2 Xbar @ under floor long. load
c (mm) long.		plenum: dash to w/s
h (mm) vert.		roof to rr comp pan
m (mm) long.		rrend pnl to rr long rail susp load pt
tf (mm) lat.		span between CL's of frt strut caps lt & rt
P3 (N) vert.		fod @ underfloor long. Load
P8 (N) vert.		rr long rail @ 6 Xbar load
P6 (N) vert.		2 Xbar @ rkr inr load
P1 & P5 (N) vert.		upr rail @ strut tower & FBHP inr loads
e (mm) lat.		CL strut cap to inner tower
f (mm) lat.		shock cap width
u (mm) lat.		CL strut cap to upr rail inr

Figure 12.11 Data form.

MATRIX EQUATION PART NAME	SOLVED Q EDGE FORCE NAME	EDGE FORCE
floorpan	cross-bar lat.	Q1
rr bulkhead	rr blkhd/cab side vert	Q2
frt bulkhead	dash/cab side vert	Q3
rr wdo frm	rr wdo frm side vert	Q4
w/s frame	w/s frm/cab side local	Q5
roof	roof/cab side long.	Q6
cab side	floor/cab side long.	Q7

Matrix for coefficients:

	1	2	3	4	5	6	7'
1	-L	0	0	0	0	0	B
2	-h4	-B	0	0	0	0	0
3	-h1	0	B	0	0	0	0
4	-h3	0	0	B	0	0	0
5	h2	0	0	0	B	0	0
6	I	0	0	0	0	-B	0
7	0	L	0	-L	0	H-h1	h1

name	value	definition
L (mm) long		cab length
B (mm) lat		cab body width @ rkr flange
h1 (mm) vert		FOD height
h2 (mm) true		w/s length
h3(mm) vert		rr wdo frm height
h4 (mm) vert		cab rr blkhd height lower
I (mm) long		roof length
H (mm) vert		cab height
h (mm) vert		cargo box height
L3 (mm) long		cab rr blkhd to rr susp
L1 (mm) long		FOD to frt susp

name	value	definition
P1 (N) vert		FOD @ midrail vert
P3 (N) vert		2 Xbar @ long. vert
Sp (mm) lat		CL frt strut span lt-to-rt
Qr (N)		n/a
Qx2 (N)		n/a
Rr (N) vert		rear susp input @ rail
Sr (mm) lat		CL rr susp spring lt-to-rt
Rp (N) vert		frt susp unit load input@ strut
Rp' (N) vert		frt susp input @ midrail
Moment (N-mm) X-Z		SV moment: frt strut to sfrm
Ru (N) vert		frt susp input @ upr rail

name	value (SOLVED Q EDGE FORCE (N))	definition (CALCULATED INTERNAL LOADS)
e (mm) lat		CL strut to inner tower lat
f (mm) lat		shk tower width
u (mm) lat		CL strut to upr rail outer lat
a1 (mm) long		FOD to #2 Xbar
a2 (mm) long		cab rr blkd to #4 Xbar
P4 (N) vert		2 Xbar @ rkr inr
xf (mm)		n/a
P5 (N) vert		4 bar @ rr long rail
P6 (N) vert		cab rr blkd @ rr long rail
P7 (N) vert		4 Xbar @ rkr inr
xr (mm) lat		rr long. rail to rkr inr lat

Figure 12.12 Data form.

Step 5 Arrange the results in tabular form to compare the forces to the baseline model. From the set of input parameters selected for this problem, the results for steps 3, 4, and 5 are shown in Figure 12.13.

Step 6 Analyse the results. The priority is to consider the forces on the platform members first, since the business case was based on retaining a large percentage of common parts. The forces affecting the common platform parts are shaded in the left-hand column of Figure 12.13. Percent increases greater than or equal to 10 per cent are also indicated in the table:

- Q Cross-Car Edge Lateral, Q Floorpan/Sideframe Edge Fore-Aft, P #4 Xbar @ rr longit. rail load vertical, P #4 Xbar @ sideframe load vertical, and P rr bulkhead lower @ rr longit. rail load vertical forces are especially impacted by the Pick-Up Truck variant.
- The Q Rear Comp Pan/Sideframe Edge Fore-Aft and P rear long. rail @ #6 Xbar load vertical forces are especially impacted by the Mid-Size Van variant.
- The Q FBHP/Sideframe Edge Vertical force is impacted by both the Mid-Size van and (more favourably) by the Pick-Up Truck.
- The Sideview Bending Moment-Upper MC Rail @ FBHP, P #2 Xbar @ under floor longit. load vertical, and P #2 Xbar @ sideframe load vertical forces are impacted by all model variants because of the increased moment arm between the dash plane and the front wheels.

The above observations are the first indication that the original assumption of '*carry-over*' may not be feasible for some of the platform structural members. Each case will need to be analysed to assess the impact on structural member size and mass. In the following analysis it is assumed, unless otherwise noted, that *stiffness* is the governing criteria. That is, the stiffness of the *baseline* model members are considered to be acceptable and serve as the target. Therefore, using unit loads as input and the deflection results for comparison are appropriate. It must be borne in mind, however, that the effect of the increased weight of the other models (and therefore higher suspension loads) has not yet been factored into the calculations.

Dash (front bulkhead panel)

The FBHP-to-sideframe vertical edge force is an order of magnitude less than the other edge forces. Furthermore, the pick-up truck edge force is significantly lower than the other variants. This is because the lateral edge force is higher, thus the amount of additional force required to balance the forces generated by the midrails is reduced. The 30 per cent higher lateral edge forces at the base of the windshield and the front of the floorpan translate into higher shear flow q which must be managed. This is further explained in the next section.

Cross-car lateral and floorpan-to-sideframe fore–aft edge loads

The pick-up truck structural arrangement has caused these forces to increase by over 30 per cent. This in turn translates into an increase in the shear flow q, which equals Q/unit

Force Name/Type Q, P (N) M (N-mm)	Baseline Production Mini-Van	Station Wagon	% Change	Mid-Size Van	% Change	Crew Cab Pick-Up Truck	% Change
Q Cross-Car Edge Lateral	518	546	5	531	2	695	34
Q Cowl/FBHP Edge Fore-Aft	110	116	5	113	2		
Q A-Pillar/Sideframe Edge Local	233	224	4	217	7	416	79
Q Roof Side/Sideframe Edge Fore-Aft	994	877	12	1051	6	977	2
Q Floorpan/Sideframe Edge Fore-Aft	787	829	5	806	2	1051	34
Q Rear Seat Cross-Bar/Sideframe Vertical	458	463	1	439	4		
Q Rear Lift-Gate Opening/Sideframe Edge Vertical	31	43	41	8	73		
Q Cab Rear Window/Mid-Gate Upper to Sideframe Edge Vertical						269	
Q Cab Rear Bulkhead/Mid-Gate Lower to Sideframe Edge Vertical						1238	
Q Rear Comp Pan/Sideframe Edge Fore–Aft	481	507	5	540	12		
Q FBHP/Sideframe Edge Vertical	67	69	3	76	14	2	97
M, Sideview Bending Moment - Upper MC Rail @ FBHP	133 158	167 465	26	167 465	26	167 465	26
P upr rail @ strut tower & FBHP inr load vertical	579	578	0	578	0	577	0
P strut cap load @ midrail load vertical	421	423	0	423	0	423	0
P fod @ underfloor longit. load vertical	534	565	6	565	6	565	6
P #2 Xbar @ under floor longit. load vertical	113	142	26	142	26	142	26
P #2 Xbar @ sideframe load vertical	59	79	34	79	34	79	34
P #4 Xbar @ rr longit. rail load vertical						985	
P #4 Xbar @ sideframe load vertical						739	
P rr bulkhead lower @ rr longit. rail load vertical						2068	
P rear seat Xbar @ rr longit. rail load vertical	693	704	2	670	3		
P rear long rail @ 6 Xbar load vertical	374	379	1	414	11		

KEY

> 10% force increase to proposed common platform parts	34
> 10% force increase to proposed new unique parts	79
% force decrease	2

Figure 12.13 Force comparison.

length. The floorpan-to-sideframe q is also higher because its length is shorter. These higher loads must be transferred by the spot welds. The spot weld forces should be checked to determine, especially at stress-raiser locations, if the number of spots is still adequate. A preliminary investigation could be performed with the aid of an existing shell finite element model of the baseline van body-in-white. The maximum expected torsion load would be applied per estimates provided in Chapter 2, section 2.4.2. The resulting forces at the spot-welds and the nearby nominal stress values would be increased by 30 per cent and checked against criteria, plus any appropriate safety factor. It should be noted that the higher cross-car lateral edge force also has ramifications for the roof pillars, especially at the *windshield frame*. It was assumed that the windshield frame would be a new part, but that the cross-sectional dimension criteria (moment of inertia I) could be carried over. Given similar dimensions, the moment on the upper and lower corners of the pick-up truck windshield frame in the $Y-Z$ plane also increases by 30 per cent relative to the other model variants. Idealizing the windshield frame as a ring frame (Figure 5.16) shows it to be in complementary shear and that the maximum bending moment in the corners is $QD/4$, where D is the height of the frame. The stiffness K in the direction of the lateral edge force Q (from Chapter 7, section 7.2.4) is equal to $(24E)/\{D^2(D/I_2 + B/I_1)\}$. The corresponding deflection is equal to $Q\{D^2(D/I_2 + B/I_1)\}/(24E)$. The deflection will increase in proportion to the increase in Q, given similar height D. If the target is to maintain the same structural rigidity then I_1 and/or I_2 will need to increase, all other factors assumed to remain unchanged. This will translate into either an increase in the cross-sectional dimensions and/or an increase in part thickness for the windshield frame structural members. The above equation can be manipulated to estimate the amount of increase required in either I_1 and/or I_2.

Motor compartment upper rail @ FBHP

The sideview bending moment has increased due to the specified forward movement of the front wheels relative to the front-of-dash plane. It was assumed originally that, except for length, the design and most of the associated parts could be carried over from the baseline van. However, the higher moment requires a cross-sectional increase in order to maintain the bending stiffness. Idealizing this situation as a cantilever beam, the vertical deflection at the suspension load point is $PL^3/3EI$ (Blodgett 1963). For equal deflections this means:

$$PL_{\text{base}}{}^3/3EI_{\text{base}} = PL_{\text{new}}{}^3/3EI_{\text{new}}$$

and therefore the ratio $\qquad\qquad I_{\text{new}}/I_{\text{base}} = L_{\text{new}}{}^3/L_{\text{base}}{}^3$

The I_{new} property will translate into greater cross-sectional dimensions and/or increase in part thickness, or perhaps an additional reinforcement.

#2 Crossbar

The vertical forces on the member from the underfloor longitudinal rails are increased by 26 per cent and the sideframe end forces by 34 per cent as a consequence of the

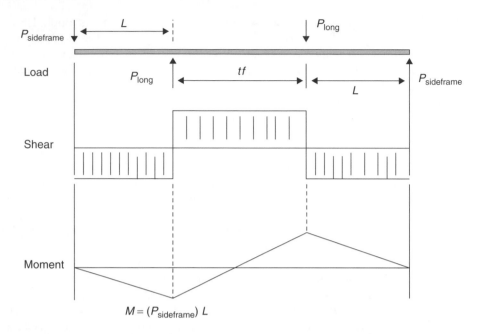

Figure 12.14 #2 floor cross-member load, shear and bending moments.

front wheel forward movement. The shear and bending moment diagram is shown in Figure 12.14.

The deflection at the longitudinal rail interface where the maximum bending moment occurs will be increased proportionately with the increase in bending moment, since the cross-beam dimensions and E remain constant. The deflection equation will have the product EI in the denominator. To maintain the same deflection, the I inertia property will need to increase by the ratio $I_{new}/I_{base} = (P_{sideframe\ new}/P_{sideframe\ base})$. Again, this implies a proportionate increase in crossbar section dimensions and/or part thickness.

Rear longitudinal rails

It has already been mentioned that the structural load path for the pick-up truck is different from the other model variants in how the rear suspension vertical forces are reacted. It has been initially assumed that this member would be essentially a common and carryover part, except for adjustments in length. However, the pick-up truck arrangement causes high loads to be transferred from the rails to the cab rear bulkhead and the #4 crossbar. By contrast the rear longitudinal rail loads in the van and station wagon are more evenly distributed. Recalling the differences between the two arrangements shown in Figures 12.7 and 12.8, the translation of these into loading diagrams is shown in Figures 12.15 and 12.16.

It is obvious that the loading conditions on the longitudinal rail are not the same between the van/station wagon and the pick-up truck. This in turn translates into differences in bending moments shown in 12.17.

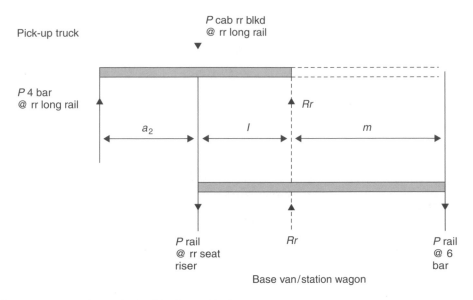

Figure 12.15 Rear longitudinal rail loading – side view.

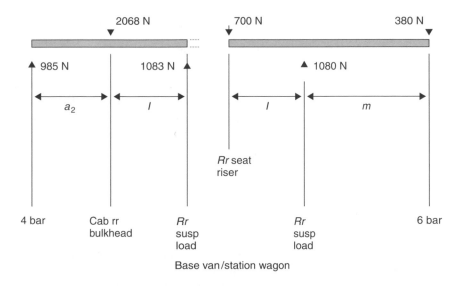

Figure 12.16 Rear longitudinal rail loading – side view.

Note from Figure 12.17 that the maximum bending moment on the pick-up truck rail is about double that of the van/station wagon. For the van/station wagon the vertical deflection at the rear suspension load point is:

$$Rrl^2m^2/3EI_{\text{van/station wagon}}L \quad \text{(Blodgett 1963)}$$

where $L = l + m$.

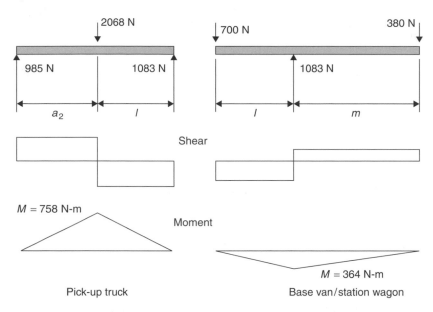

Figure 12.17 Rear longitudinal rail loading – side view.

For the pick-up truck the vertical deflection at the rear suspension load point is:

$$(Rrl^2/3EI_{\text{truck}})^*(a2+l) \quad \text{(Blodgett 1963)}$$

Equating the deflections:

$$\cancel{Rr}\,l^2m^2/\cancel{3}\cancel{E}I_{\text{van/station wagon}}L = (\cancel{Rr}\,l^2/\cancel{3}\cancel{E}I_{\text{truck}})^*(a2+l)$$

$$m^2/I_{\text{van/station wagon}}L = (1/I_{\text{truck}})^*(a2+l)$$

$$I_{\text{truck}}/I_{\text{van/station wagon}} = a2 + l(l+m)/m^2$$

Substituting the respective values yields $I_{\text{truck}}/I_{\text{van/station wagon}} = 2.34$.

This indicates that, if the same stiffness is to be maintained, the pick-up truck rear longitudinal rails will require an increase in cross-section size and/or thickness over the van/station wagon.

#4 Crossbar

On the van and station wagon, this member is sized to support local chassis component attachments but does not help to support the rear suspension load reactions. On the pick-up truck, it is an essential element to supporting the rear suspension loads. This would warrant an analysis similar to the one performed on the rear longitudinal rails if it is originally assumed that the part is common.

The above studies for the torsion load case have suggested that there are some flaws in the original assumptions concerning common and carryover parts to support the business case. The bending and mass load conditions also have yet to be studied, and will add other issues that need to be considered.

In this above hypothetical case study, it will be decided to retain the prescribed forward movement of the front wheels or else the new engine will not fit. The new engine is deemed sufficiently important for the product's positioning that it will offset the investment required to change the motor compartment upper rail and FBHP tooling. However, the rear longitudinal rails are another matter. The effect of the pick-up truck on the rear longitudinal rail commonality is deemed unacceptable. It is too expensive to tool unique parts for the pick-up truck. Sizing the rails of all the models around the pick-up truck would solve the commonality issue but adds cost and weight to the station wagon and van that are not acceptable.

The pick-up truck configuration also puts higher loads on the cab body, compared to the van and station wagon. The higher edge forces indicate that additional spot welds may be required which were not accounted for in the manufacturing plan. Though it was assumed to be a new part anyway, the additional structural stiffness that may be required for the windshield frame was also not considered in the original business case. Overall, the pick-up truck is considered to be a highly desirable model variant but its viability is at risk unless solutions are found for the above issues. The free body diagrams are revisited to explore alternative load paths. The main problem stems from the loads transmitted to the cab from the rear longitudinal rails, and the manner in which they are supported. A means must be found either to reduce the input loads or to redistribute them in such a manner that the forces on the longitudinal rails are reduced.

Fundamentally, there are two potential paths for the rear suspension load to be introduced to the cab: (a) from the longitudinal rails (previously discussed) or (b) from the cargo box side member. In the statically determinate system that is assumed, it must be one or the other. In reality, the forces will be distributed according to the relative stiffness of the two load paths. The selection of a structural design strategy will determine how the load paths will be biased. Finite element analysis would be used to develop and verify that the desired load path distribution has been achieved.

Earlier free body diagrams of a pick-up truck shown in Figure 6.26 indicate that distributing load to the cargo box side panel will result in horizontal loads on the cab sideframe that in turn will produce bending moments on the B-pillar. These bending moments will need to be comprehended in the sizing of the B-pillar. Because the business case originally assumed that the cab sideframe and B-pillar would be new anyway, there is expected to be relatively minimal cost impact if these parts need to be modified to accommodate higher loads. In order to transfer more force to the cargo box side, there needs to be a sufficiently stiff load path from the rear longitudinal rail to the inner wheelhouse. This could be similar to what was illustrated at the bottom of Figure 10.3. In this particular case the rear longitudinal rail cannot be simply moved outboard due to part commonality and manufacturing considerations in body assembly. However, the spring seat and jounce bumper can be moved outboard so that more load is distributed to the outer cargo box through the wheelhouse inner panel. An intermediate stiffener is necessary to ensure that the spring seat does not collapse under extreme suspension loads, so a gusset is employed. This design proposal is illustrated in Figure 12.18.

The first finite element analysis of this proposal determines that the load transfer from the rail to the wheelhouse is insufficient, despite the spring seat being moved outboard. A second design alternative at the bottom of Figure 12.19 provides a second gusset that helps to transfer load from the rail vertical side wall into the wheelhouse inner panel.

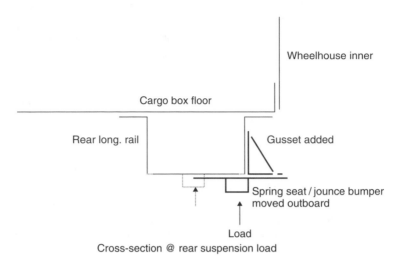

Figure 12.18 First alternative load path for rear suspension load.

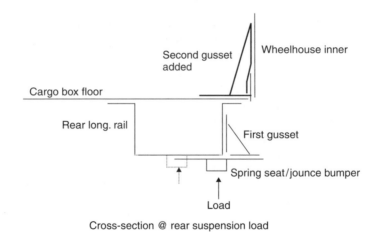

Figure 12.19 Second alternative load path for rear suspension load.

Additional bracing may need to be provided between the wheelhouse and cargo box side panels to ensure sufficient load transfer. Recall that the real problem is statically *in*determinate, so it will be necessary to use finite element analysis to verify that the load is being distributed as intended.

Although redistributing load into the B-pillar is acceptable from a part commonality standpoint, the section size will need to be rechecked to accommodate the higher loads. A coarse finite element model can verify the estimated forces going into the B-pillar with the new modification. These forces can be used to recalculate the required section properties using basic engineering formulae. The finite element model can then be subsequently employed to optimize the member properties.

The above study will need to be repeated for the bending and durability cases. Here, as illustrated in previous chapters, the edge force calculations are not as numerous as in the torsion case. As mentioned earlier, the higher mass of the other model variants will also need to be factored into the calculations when compared to the baseline van, especially in cases where durability (extreme load capacity) is the governing design constraint. The durability load case should be supplemented with a conceptual finite element model to check for local stresses at critical points, such as where the pick-up truck cargo box connects to the cab.

Of course, other factors will also need to be considered which are not developed in this section: crashworthiness, noise and vibration, fatigue, manufacturing and assembly. The important thing to recall from this chapter is the process of applying SSSs and engineering fundamentals early and rapidly to assess the impacts of various alternatives. Free body diagrams can be generated in minutes to gain a qualitative understanding of the issues involved, prior to the development of computer generated design data and finite element models. The formulation and calculation of the edge forces and internal reactions can be accomplished within hours to gain a comparative assessment with minimal computational resources. Significantly less time will be required if SSS models have been previously developed for different vehicles and structural arrangements which can then be applied off-the-shelf. These results can then be used to develop the structural load–path strategy and guide the course of finite element studies.

References

Bastow D., Car Chassis Frame Design, *Proc. IAE*, Vol. XL, p. 154, 1945

Bastow D., *W.O. Bentley, Engineer*, Haynes Publishers, 1978

Beermann H.J., (Translation edited by G.H. Tidbury) *Analysis of Commercial Vehicle Structures*, Mechanical Engineering Publications, 1989

Blodgett O.W., *Design of Weldments*, James F. Lincoln Arc Welding Foundation, 1963

Booth A.G., Factory Experimental Work and its Equipment, *Proc. IAE*, Vol. XXXIII, pp. 503–546, 1938

Costin M., Phipps D., *Racing and Sports Car Chassis Design*, Batsford, 1961

Den Hartog J.P., *Strength of Materials*, McGraw-Hill, 1949

Donkin C.T.B., *The Elements of Motor Vehicle Design*, Oxford University Press, 1926

Erz K., Uber durch Unebenheiten...... Automtechn. Zeitschr., 1957

Garrett T.K., Automobile Dynamic Loads, *Automobile Engineer*, pp. 60–64, Feb., 1953

Garrett T.K., Structural Design, Part 1: An Analytical Method for Chassisless, Vehicle Design *Automobile Engineer*, March, 1953

Gere J.M. and Timoshenko S.P., *Mechanics of Materials*, 3rd SI Edition, Chapman and Hall, 1999

Megson T.H.G., *Aircraft Structures for Engineering Students*, Arnold, 1999

Nardini D., Seeds A., Structural Design Considerations for Bonded Aluminium Structured Vehicles, *SAE Technical Paper 890716*, 1989

Pawlowski J., *Vehicle Body Engineering*, Business Books, 1969

Pawlowski J., *Vehicle Structures*, Cranfield University, 1986

Perry D., Azar J.J., *Aircraft Structures*, 2nd edn, McGraw-Hill

Roots M., Brown J.C., Anderson N., Wanke T., Gadola M., The Contribution of Passenger Safety Measures to Structural Performance in Sports Racing, MSC World Users Conference, Newport Beach Ca., 1995

Ryder G.H., *Strength of Materials*, MacMillan, 1969

Shigley J.E., *Mechanical Engineering Design*, 1972

Swallow W., Unification of Body and Chassis Frame, *Proc. IAE*, Vol. XXXIII, pp. 431–475, 1938

Tidbury G.H., Measurement of Loads Between the Suspension and the Body Structure of a Small Car, Conference on Stresses in Service, Institution of Civil Engineers, London, 1960

Ultralight Auto Steel Body, Final Report, AISI, Washington DC, 1998

AISI, Maple V Software

Automotive Engineering magazine, Vol. 92, No. 9, pp. 79–80, Sept., 1984

EDSU, Buckling in Compression of Sheets between Rivets, Engineering Sciences Data Unit, Data Sheet 02.01.08

EDSU, Buckling Stress Coefficients for Curved Plates in Shear, Engineering Services Data Unit, Data Sheet 02.03.18/19

EDSU, Buckling of Flat Plates in Shear, Engineering Services Data Unit, Data Sheet 71005

EDSU, Natural Frequencies of Rectangular Flat Plates with Various Edge Conditions, Engineering Services Data Unit, Data Sheet 75030

Index